The Atheist's Guide to Christmas

The Atheist's Guide to Christmas

Edited by
ROBIN HARVIE
and
STEPHANIE MEYERS

HARPER PERENNIAL

NEW YORK • LONDON • TORONTO • SYDNEY • NEW DELHI • AUCKLAND

HARPER PERENNIAL

HarperCollins books may be purchased for educational, business, or sales promotional use. For information, please write: Special Markets Department, HarperCollins Publishers, 10 East 53rd Street, New York, NY 10022.

FIRST U.S. EDITION PUBLISHED 2010.

Library of Congress Cataloging-in-Publication data is available upon request.

ISBN 978-0-06-199797-6

10 11 12 13 14 OV/RRD 10 9 8 7 6 5 4 3 2 1

CONTENTS

ARTS

EVENTS

The Atheist's Guide to Christmas

WELCOME

Welcome to *The Atheist's Guide to Christmas*, the atheist book it's safe to leave around your grandmother. Here, you'll find no chapters titled "666 Ways to Diss the Pope," "A Beginner's Guide to Church Graffiti," or "How to Bash the Bishop."

What you will find are brilliant contributions from forty-two* of the world's most entertaining atheist scientists, comedians, philosophers, and writers, who have all donated their time, thought, and jokes for free to help you enjoy Christmas.

Maybe you bought this book for yourself, or perhaps there's a price sticker over the *A* of *Atheist* and your devout great-aunt bought it for you, hoping to make you more religious. Either way, all royalties are going straight to the UK's leading HIV and sexual health charity, Terrence Higgins Trust, so to whoever bought it: thank you. (What do you mean, you haven't bought it yet and you're still loitering in the bookstore reading this with your grubby thumbs on the pages? Take it to the counter this instant!)

We hope you enjoy every page, and that you have a truly excellent Christmas.

---

* Because forty-two, as explained in *The Hitchhiker's Guide to the Galaxy*, is the answer to life, the universe, and everything. If you haven't yet read it, you might want to buy it along with this book (although the salesclerk may then think you only buy books with the title format *The* ____ *Guide to* ____).

STORIES

Truth is more of a stranger than fiction.

—Mark Twain

It's Beginning to Feel a Lot Like Christmas

Ed Byrne

"I've already done all my Christmas shopping for this year. I bought all my aunties socks and Y-fronts. See how they like it."

For many years, that was my only Christmas joke. Seeing as Christmas can be quite a lucrative time for a jobbing comic, a time when you can get paid two or even three times your normal fee in compensation for having to entertain people who are two or even three times more drunk and rowdy than normal, you would think I would have written a slew of seasonal zingers to keep the paper-hatted hordes chuckling into their lukewarm mulled wine. But I never did. I would kick off with my little morsel of Christmas hum-buggery and then carry straight on with my usual cavalcade of jokes about smoking, drinking and slagging off Alanis Morissette. Why, I imagine you're wondering, was this so? Why would somebody who, particularly in his early circuit days, was so eager to churn out crowd-pleasing material not hit that stage with an arsenal of Yuletide yuk-yuks? Surely someone with such a pragmatic approach to comedy would have at least a solid five minutes of holiday-based lateral thinking thrown into a box of sarcasm, wrapped in whimsy paper, and all tied up in the pink bow of impeccable timing. But no.

The reason for this is simple: I have always found it easier to write jokes about things I hate, and I don't hate Christmas. Sure, there's been some dodgy stuff left for me under the tree over the years. "Oh, did Santa run out of Scalextric sets? Well, I suppose Tamyanto make one just as good." The Santa Claus that came to our house did not believe in paying for advertising. As I grew older and Santa was replaced by my parents, they continued in this vein. Maybe they were early anti-globalization activists and thought they should boycott major bicycle manufacturers like Raleigh or Dawes. Maybe that's why at the age of fourteen I was the proud owner of the only Orbita ten-speed in all of North County Dublin.

It wasn't that my folks were being cheap. They were just doing their bit to fight the power of Big Bike. I'm not saying that Orbita don't make a quality product, but I can't help but think that they could have built up much better word of mouth if they hadn't sold my dad a bike with two right pedals. Yes. Two right pedals. When it comes to bicycle pedals, two rights make a wrong. He did try to return the bike a couple of days later, but found out the hard way that a gift shop that wasn't there before December 1 won't be there after December 24. Well, I say *he* found out the hard way. He wasn't the one pedaling to school with only one foot. By the time I was fourteen, I was so asymmetrically developed it took all my concentration not to walk in a circle.

Crappy presents notwithstanding, I've always been a big Christmas mush, enjoying the sentimentality of the season. New Year, I've always felt, can go and shite. Maybe that's because as a kid I always used to babysit the neighbors' kids so that the neighbors could go to a party at my parents' house. But Christmas has always been my favorite time of the year. Even going to mass—a pastime I obviously have little love for if I'm included in this book—was more fun on Christmas Day because we all got to look at each other in our Christmas clothes. Those of us who got decent trendy-looking ones got to point and laugh uproariously at the chunky-knit efforts of those less fortunate. This was one aspect of Christmas where my mother never let me down. We couldn't afford Armani, but at least I never had to endure the humiliation of a reindeer on my sweater at age thirteen.

So Christmas has always been in my cool book. I've always found it easier to make fun of holidays like Halloween, which must be a very difficult time for pedophiles who are really trying to shake the habit. Imagine! You've got the urges. You know it's wrong, so you lock yourself in the house out of harm's way. October 31 rolls around and kids are knocking the door down. All of them dressed in cute little outfits, asking for candy. You don't even have to offer. Sweets are being requested. That's almost entrapment, if you ask me.

However, much like everything else since I hit my thirties, certain things are beginning to annoy me about my favorite holiday. Sure, there are the usual headaches that just come as you get older. Not enough time to go shopping. Swearing that next year you won't leave it too late to do it online. Trying to come to a compromise with your wife regarding whose family you should spend it with—yours, hers, or perhaps some neutral family that you both loathe equally. Everything gets more complicated as you get older, and the responsibilities of adulthood are always going to do their best to choke the living joy out of any occasion. I'm not really talking about that. I'm talking about something that I used to find exciting about Christmas as a youngster but as an older man I just find wearisome, and that is the length of the lead-up to it.

As you get older there are three things you observe: policemen are getting younger, teenage girls are dressing more like prostitutes, and Christmas comes earlier every year.

Christmas is a special time for a lot of us, and the rituals, sights, smells, and sounds that go along with it can be very effective at stirring up childhood memories of Christmases past and generating a nostalgic, sentimental glow. But if shops start hanging tinsel in October, it doesn't take long for the spell to be broken. Seriously: when you hear Wizzard's "I Wish it Could be Christmas Everyday," does it remind you of sipping mulled wine next to a roaring fire or does it remind you of November in Woolworth's?

I was in my local Tesco a couple of years ago and they were selling Christmas food *in September*. That's too early. Mid-September and they had shelves of stollen, Christmas pud, and mince pies. Nobody is so organized that they buy food three and a half months in advance. Anyone who is that organized makes their own food. Just out of curiosity I pulled a pack of mince pies off the shelf to check the "best before" date, and I swear to you it was November 10. What sort of numpty buys mince pies that go bad in November? And don't tell me that some people might just want to eat mince pies in September. You only eat mince pies at Christmas, and most of us don't even like them then. I guess the logic is, they're generally so foul you can't tell if they've gone off or not. Personally, I think you may as well wipe your arse on some digestive biscuits and hand them round as shove a mince pie under my nose, regardless where we are relative to its "best before" date.

What nearly made my wife and I weep genuine tears of actual sadness was the fact that they were also selling single slices of Christmas cake. Imagine that. Not two slices, maybe for a couple who couldn't be bothered to make a whole cake. No. One slice. That's a slice for you and no slice for your no pals. It's important, now and again, to spare a thought for those less fortunate than us who might be spending Christmas alone, but I don't need such a stark reminder as single slices of Christmas cake on sale in September. That means that, with over three months to go, the bloke in question is already resigned to the fact that he'll be on his tod this festive season. He's already got it all planned out. "I'll have a Bernard Matthews Turkey Drummer, followed by a single slice of Christmas cake. Then I'll open the card I sent to myself. After which I'll stand on one end of a cracker and pull the other, get drunk, have a wank under the mistletoe, and pass out. Happy holidays!"

As depressing a notion as that is, is it any more depressing than the thought of somebody buying mince pies that go bad in November? Because, for me, that conjures up images of people who, for some reason, have had to have Christmas early this year. Nobody has an early Christmas for a happy reason. It's more likely to be a sad reason, like "Granddad's not going to make it to December. We're having Christmas in November this year and we're going to enjoy it! We'll tell him it's December. He's so far gone he won't know the difference." Either that or "We have Christmas in October so that Uncle Brendan can spend it with us. He generally goes back to prison shortly after that. It's not really his fault. He does try to stay out of trouble, but he tends to fall off the wagon at Halloween."

(Do you see what I did there? That was called reincorporation. It's a classic comedy trick. You probably thought it was strange that I should even have mentioned Halloween in an essay about Christmas, initially. You probably thought I was just padding out my piece with a bit of Halloween filler. But I wasn't. All the while I was building to that Uncle Brendan callback. Pretty clever, huh?)

So, what am I trying to say here? I guess the point I'm making is that shitty Scalextric knockoffs and bikes with two right pedals didn't dampen my enthusiasm for Christmas, but greedy retailers who try to get me into a premature Christmas mood do. I propose a moratorium on all kinds of Christmas marketing before mid-November. The Advertising Standards Authority should introduce a rule saying sleigh bells may not feature in ads until the first week of December. And while we're at it, let's introduce a law banning the sale or display of tinsel in shops until December 15. Failing that, I think Wizzard should get back in the studio and record a song called "It Should Only Feel Like Christmas One Month a Year."

Revenge of the Christmas Spirit

Neal Pollack

Around the time that my son opened a gift to reveal a SpongeBob SquarePants edition of Connect Four, bleating out, "But I already have one of these!" I reaffirmed my sacred lifelong vow to hate Christmas. Yet I'd married a woman who'd grown up celebrating the holiday, so every year it was the tree with the needles that I'd be vacuuming up until April, the endless fretting at monstrous chain stores, the baking of the Santa cookies, the putting out of the Santa cookies, the scolding the dog for eating the Santa cookies, and—my only addition to the proceedings—the watching of *Meet Me in St. Louis* on Christmas Eve, whether drunk on wine or not. And all throughout, it was Santa this, Santa that, we have to hide the goddamn gerbil so the boy thinks it's from Santa. Beyond all the terrible music and hideous sweaters and relentless spread of phony good feelings throughout the land, Santa, the cheesiest, most overcommercialized mythological figure in human history, bothered me the most about Christmas. And yet I had to *believe* in him to preserve the innocence of a child who watched Steve McQueen movies and had already called me a "fucking asshole" twice.

I suppose it could be worse. My wife could believe in *Jesus,* which she doesn't, other than giving the progressive Sunday school "I believe Jesus was a great teacher" line. I will deign to lie to my child that a fat man in a red suit comes down the chimney once a year to give him video games and Tootsie Rolls, but I won't tell him that a hippie carpenter from Israel ascended to heaven 2,000 years ago as the one true son of God.

That said, Jesus, or at least the myth of Jesus, had some fine values that didn't involve making bratty comments about cheap plastic toys on his birthday. But, it being Christmas and all, I didn't castigate my son. Instead, I took my wife into the back of the house and said, "He doesn't need all those toys."

"Leave him alone," she said. "It's Christmas."

"Yeah," I said, "and we already have a stack of unopened Lego sets from his birthday. He's never going to touch them."

From there, the conversation became serious. Suddenly I felt the immortal energy of Barack Obama's Socialist Republic of America coursing through my veins. I wanted to spread the wealth. I wanted to help people. I wanted to give those toys away.

"So, okay," Regina said. "We'll gather them up when he's not paying attention, and we'll—"

"No," I said. "I want to give them away *right now.* On Christmas."

"Are you serious, Neal?"

"Isn't that what Christmas is about? Giving?"

She sighed. I had her.

"I'll get the toys together," she said.

"I'll find a place to take them," I replied.

I went upstairs to my office and, after a quick round or five of online poker, started looking for a new home for our gifts. This was Los Angeles, after all, a land full of children, some of them quite needy. There had to be some organization to serve them, preferably one within a fifteen-to-twenty-minute drive from my house. A soup kitchen for homeless families might be nice, I thought, but those are always mobbed with rich guilty volunteers on Christmas. Then I found a nice apartment building for families where one or both parents are HIV-positive, located on Sunset Boulevard in Silver Lake, an easy two-highway drive away from my house. This seemed like a worthy cause, and a nice place for Christmas presents.

"All right," I said, coming downstairs. "I'm ready to drive. Where are my presents?"

Regina gave me the SpongeBob Connect Four and a Lego Aqua Raiders set.

"This is it?" I said.

"It's all he would give up," she said.

"You *consulted* him?"

"You can't take toys away from a kid on Christmas without asking," she said. "At least he gave something."

"True enough," I said, and then I was off to save Christmas.

One of the great joyful bonuses of Christmas in L.A., I quickly discovered, is the ability to drive without fearing for your life. I roamed from lane to lane freely, imagining that I was back in the late 1950s before this hellhole had gotten clogged with so many people, when the Dodgers were new in town and represented hope, when people looked upon rockets with wonder, where burgers were 25 cents, and when cops dumped people in ditches with no fear of reprimand. Those were the days, I thought as I reached my destination in unheard-of time.

I parked right in front, for free. O happy Christmas Day! It was a well-maintained, relatively newly built brick building on a busy commercial block. Nothing screamed "Families with AIDS live here," which was probably, I guessed, the point. I walked in; the lobby was a bit institutional-feeling, but clean and well kept. This was a place of dignity, pride, and anonymity. A foil Christmas tree, fully decorated, stood in the lobby. No presents were underneath.

"Hello?" I shouted. "Hellllllllo?"

I saw an elevator that required key access, and next to it was a locked security door. It had a buzzer, which I rang. I paced around the lobby, every so often calling out a "Helloooooo!" Briefly I contemplated leaving the gifts under the tree, like a true Santa, but then I worried that no one would see them, or that someone would see them, think they were lost, and give them to the lost-and-found. Also, I kind of wanted to get a receipt, for tax purposes.

Finally, after a while, a middle-aged black woman with a stern, social-justice-oriented gaze appeared. She introduced herself as the building's manager and asked how she could help.

"Well, I've got these extra presents for Christmas, see," I said. "And I thought they might need a good home."

She looked at me quizzically.

"Meaning, I was wondering if the kids here could use some extra Christmas presents."

Her expression said: *You dumb-ass white man, we already have presents for the kids here. Don't go taking out your liberal neuroses on us.* Then she said: "Let me see the presents."

I handed her the freshly minted junk, made by Chinese slave children, rejected by my somewhat overprivileged half-Jew spawn, and now, in the spirit of Christmas, offered to the less fortunate by a slightly hung-over unshaven forty-year-old stoner. She looked them over.

"These are pretty nice," she said.

I'm glad you approve, lady, I thought. Instead, I said, "I thought someone here could use them. Do you have any kids six and older? Those are probably the right ages. Maybe closer to eight for the Legos."

She thought about this for a minute, then said, "We have a seven-year-old and a ten-year-old. I'll see that they get them."

"Perfect," I said, as I handed them over.

"Thank you," she said.

"My pleasure, ma'am," I said, with a little tip of my Dodgers cap.

I headed for the exit.

"What's your name?" she asked.

"My name doesn't matter," I said. I was the lone bearded stranger coming into town, He-Without-a-Name, a daytime Santa Claus, the Doctor, saving the universe with a discarded Lego set.

She looked at me, hands on hips, and said sternly, "*What is your name?*"

I told her mine. She told me hers. We shook hands. I left.

Twenty minutes later I was home.

"All good?" Regina asked.

"All good," I said.

From the living room, I heard my child screaming, "This stupid stupid video game is toooooooo haaaaaaaaaaaard!"

"See what I'm dealing with?" she said.

"Predictable," I said. "I'm going to go lie down in the bedroom and watch basketball for a while."

"Anything to get out of helping around the house," she said.

"Watch it, babe," I said. "It's Christmas."

"So?"

"So I just took presents to poor children whose parents have AIDS. What have you done today?"

Her look softened. I'd gained the most important advantage you can have in any marriage, or any other relationship, for that matter: the moral high ground. I wouldn't be there long, so I was fixing to enjoy my stay. God bless us, every one!

"You're right," she said. "Go enjoy your basketball."

And that's how I discovered the true spirit of Christmas.

The Real Christmas Story

Jenny Colgan

I've always been enthralled by Christmas—the English ideal, at any rate (where I come from in Scotland, Hogmanay was always the crowd puller). The crackling snow, the animals lying down in their stalls silently at midnight in homage to the infant king, and, particularly, the glorious caroling heritage (my favorite is the rarely sung Nurse's Carol, joining the choir being the sole high point of a miserable year long ago working in a hospital):

As the evening draws on
And dark shadows alight
With slow-breathing oxen
To warm him all ni-i-ght
The prince of compassion
Concealed in a byre
Watches the rafters above him
Resplendent with fire

Good King Wenceslas, with his foreign fountains and strange ways, was as mystical to me as anything in Narnia; likewise the three kings, whose sonorous names and inexplicable gifts—

> Myrrh have I
> Its bitter perfume
> Breathes a life
> Of gathering gloom
> Sorrowing, sighing
> Bleeding, dying
> Sealed in the stone cold tomb.

—gave me strange, excited thrills.

In my teens, I dressed up as a Victorian wench and took part in carol-singing tableaux at the local castle, the same one where, years later, I would get married—at Christmastime, the pillars swathed in holly and ivy. (Incidentally, if you're having a secular service and aren't allowed to mention the word *God*, I can save you some time and effort and inform you that the only carol that legally passes muster for a non-religious Christmas wedding is "Deck the Halls.")

One of the great joys of having your own children, of course, is sharing Christmas with them. My husband, a Kiwi, spent all his childhood Christmases barbecuing on the beach and is entirely unfussed by the whole affair, but I had such wonderful Christmases that I want to make it as special as I can. Still, how to do that without fundamentally accusing their teachers of lying—or, in fact, lying?

And it is, after all, one of the greatest stories ever told—the little baby born in a manger, far from home. It has intrigue, small children (drummer boys are particularly popular in my house), stars, angels, various animals and getting to sleep outdoors—all catnip to littlies.

But, as that wonderfully conflicted cove John Betjeman put it:

> . . . is it true? For if it is. . .
> No love that in a family dwells,
> No carolling in frosty air,
> Nor all the steeple-shaking bells
> Can with this single Truth compare—
> That God was man in Palestine
> And lives today in Bread and Wine.

Because, of course, accepting the Christmas story means accepting a whole bunch of other stuff, doctrine perhaps not quite so tea-towel-and-stuffed-lamb-friendly. And now that my three-year-old is at preschool—a Catholic preschool, no less, it being our local—of course the questions have begun.

"Are you having the Baby Jesus?" he says, prodding my large pregnant stomach.

"No," I say. "That's been done."

"Oh. Are you having a monkey?"

"I hope not."

I find him in the bedroom with the lovely Nativity book his devout—and devoted—granny has sent him (even though he hasn't been baptized and thus is slightly damned and stuff), arguing with his friend Freya.

"Those are the three kings," he says solemnly.

"*No!* They're the three wise men!" said Freya, in a tone that brooks no argument.

"*No!* They are *kings*!"

"*Wise men!*"

"*Kings!*"

"*Mom! Freya says she knows my story but it is* my *story!*"

"*It is* my *story!*"

"It is," I say, "everyone's story. It is one of the most famous stories ever told. Nearly everyone you will ever meet will know a little bit about *this* story."

Wallace thinks about this for a bit.

"No. It is just mine. Grandma sent it to me."

Sometimes I feel like Charlotte in *Sex and the City*, having one last Christmas tree before she gives it all up for Judaism.

I take the boys to Christmas morning mass—where my mother is playing the organ—but they don't know when to sit or stand or what to do, and I am unaccountably nostalgic for a life I never wanted.

Christmas, for a practicing Catholic child, was seen as a reward for lots and lots and lots of church. We were constantly told that Easter was the more important festival, but Easter is, relatively speaking, rubbish. Yes, there's a chocolate egg, but six weeks of no sweets plus Stations of the Cross on Wednesdays, Good Friday mass, confession, and the Saturday vigil (*hours* long)—the trade-off is, frankly, just not worth it. Though the palms on Palm Sunday are quite good.

Christmas, on the other hand, is just normal amounts of church (except, alas, for that totally gruesome year it fell on a Saturday and we couldn't believe we had to go again the next day), but also school parties, the *Blue Peter* Advent ring, the calendar, going to Woolies to buy your mom a tiny bottle of Heather Spirit cologne (69 pence), and the glorious bellowing of "O Come, O Come, Emmanuel"—a song more than a thousand years old—all serving merely to heighten the crazed, overwhelming anticipation that could only be sated by a pack of thirty felt-tip pens, graded by shade, yellow in the middle, and getting to eat lots of very small sausages.

But there is another story too, I know, to tell my little ones; perhaps not quite as immediate, but wonderful in its own way, and it starts this way:

"In the northern parts of the world, the winters are long, and cold and dark, and people would get sad and miserable. So they have always in the very depths of winter, from the beginning of recorded time, celebrated light, life, and the promise of renewal and new birth, just when they most needed cheering up.

"And they would store food, and eat and drink and be merry. And in time different cultures and creeds passed over the world, and changed and added to the stories about why we were celebrating, and said that perhaps we were celebrating because of a green man or Mithras or Sol, or because the Baby Jesus was being born, or because Santa Claus is flying over the world—look here, NASA even tracks him by satellite (www.noradsanta.org).

"And now, like all the millions of people who lived before us, we too use midwinter to see our family and exchange gifts, and feast and be merry and carry on traditions from our ancestors."

And they will say, "Why?"

And I will say, "Because we love you."

And I will wonder, as I often do, why we love our children—our own children, not a chimera wrapped in swaddling clothes and found in a manger—so very, very much, and wishing that there were slightly more reassuring, less genetic, less cold scientific reasons that we atheists could give for why this is so.

And then I will probably just say, "Shall we sing 'Little Donkey' again?" knowing that they will immediately rush off to fetch their sweet Christmas bells.

A Child Was Born on Christmas Day

Emery Emery

Being born on December 25, I often found myself quite melancholy around the holidays. When I was a child it was simply not possible for my family to give me the special attention that most enjoy on the hallowed day of their birth. For children unfortunate enough to share their birthday with Jesus, Christmas is an unholy day of disappointment and loneliness.

Every other birthday party I attended was clearly a day set aside specifically to celebrate one person's most important life event: emerging from deep within his or her mother's womb and surviving the ordeal. I had survived, but as it turns out, the Christians believe that Jesus was born of a virgin on December 25, and they deem it a miracle. How can any kid compete with that?

My grandmother raised me for my first ten years, and she tried her best to make me feel special every Christmas. She would bake a cake just for me. One year it was in the shape of a snowman, and another it was Santa's face. I especially enjoyed the Santa cake because I was allowed to take a knife to good ol' Saint Nick. There was a cathartic quality to it. I don't remember any Jesus cakes, but that would have been nice as well.

Even though Grandma tried to make Christmas just a bit more about me, her efforts always fell short as throngs of family poured into the house to exchange gifts with each other and give me my two-birds-with-one-stone presents. "Happy Birthday and Merry Christmas" was often written on the gift tags. I recall plotting to give people birthday gifts that said "Happy Birthday and Merry Christmas," and I would then make a conscious decision to not give them anything on Christmas Day. But somehow I just couldn't go through with it.

During one of my early teenage years, in a conciliatory effort, my mother decided my birthday would be celebrated on the half year, June 25. I thought this was a really great idea, and I was insanely excited. I ran to my room and marked it on my calendar. Sadly, Mom was not very good with follow-through, and while she may well have marked a calendar herself, she had forgotten to check it. June came and went without any fanfare. Needless to say, my disappointment grew even more profound.

Every year that passed brought another Christmas that left not just me unfulfilled but my sister as well. Unfortunately, she had been born on Christmas Eve, one day short of a year after I was born. Just as I suffered the unfortunate side effects of being swept aside to make room for a grand celebration of the birth of the Baby Jesus, my sister endured the same profound injustice. Not only would our day not be ours, it would be everyone's. Both my sister and I had to split what tiny amount of birthday we were able to cobble together.

One particularly lamentable Christmas, my sister received two identically wrapped packages from our mother. She unwrapped one to find a single, fairly cheap earring. As she unwrapped the other box, revealing the matching earring, Mom exclaimed, "One is for your birthday, and the other is for Christmas!" I wish I could report that my sister let loose with an impressively long string of absurdly creative expletives, but I have no memory of this particular event. I suspect I was sitting quietly next to the tree attacking the manger with GI Joe, a common, seasonal practice of mine.

One year, according to my mother, she had done everything she could to give us a classic birthday. She had planned a huge party for my sister and me. She invited all our friends and scheduled the party for December 23, which fell on a Saturday that year. While not all my friends were able to be there, with holiday travels and family gatherings preempting our party, many of our friends were indeed present, and I am told we had a great birthday party.

While I have no doubt that my mother remembers it that way, I do not have any memory of this amazing party. Any psychologist worth his or her weight in Freudian dogma may be able to explain why I would have no memory of it or why my mother would remember it so clearly, but what I know for sure is that I have no recollection of any Christmas that is fond. This party may have happened and my mother may have had an amazing time, but I was not present at any such event.

Through most of my childhood, I wished Christmas didn't exist, and I harbored ill will toward all who enjoyed it. It made me angry and sad. I felt that I was being robbed by Jesus, Santa, all the reindeer, and everyone I knew. Then, as a young adult, I found myself investigating Christmas, and discovered some interesting information.

While no one seems to agree on the actual day of Jesus' birth, most scholars agree that it wasn't December 25. Some have it in November. Others claim it was in March, and still more believe it must have been in September. But whatever day it was, it clearly wasn't on my birthday, and that makes it even worse. Here I am, being robbed of my very own day by a ritual that isn't even accurate! If only there were a God to pray to and ask for some kind of retribution.

The history of the day of my birth is tainted by an unthinkable practice, and here I am in the twenty-first century, feeling slighted and sad.

My point is this: any child born on Christmas cannot have a real birthday. It's not possible. There are some who have claimed that I turned to atheism due to my birthday melancholy, but while I will never celebrate my day of birth on the level that most enjoy theirs, I am not an atheist because of this. I am an atheist because I reject all stories that are not rooted in and supported by empirical data—because I do not need to have stories that make me feel better about that which I do not know or that which I fear.

Now, as a full-grown adult with my destiny in my hands, I hold myself responsible for my own happiness and no longer sit around, sullen and depressed, every Christmas. In fact, I enjoy celebrating Christmas in my own way. My wife and I fly out to visit her parents each year—usually on Christmas Day, in fact. Since most people think the day sacred, flights are usually half price, and if they're overbooked, we often give up our seats in exchange for travel vouchers. One Christmas evening we did just this, had a lovely evening in a nice hotel, got up on the twenty-sixth, flew into our destination, and had a wonderful dinner with my wife's parents. We awoke on the twenty-seventh, had a very nice gift exchange, ate birthday cake, and played in the winter snow. While my wife's parents believe in God, they aren't really much for ritual. They just look forward to seeing us for the holidays, whichever day we arrive.

Whether we're traveling, staying in a hotel, or enjoying my wife's family, December 25 isn't Christmas Day to us. My wife has taken to referring to it as "Emerymas." Sure, Emerymas is a contrived and fully invented construct meant to mark the birth of my wife's husband. But why not? If ancient priests could do it, so can my wife.

If you're a kid born on the twenty-fifth, Christmas sucks. Emerymas, however? A day like any other day, with one very distinct exception: I was born. And according to my wife, that's something to celebrate.

I appreciate all that my mother and my grandmother tried to do. They can't be held responsible for my failed childhood birthdays—they were up against eons of ritual and tradition. Still, if I'd been alive in the fourth century, I could have been sacrificed by pagans, so perhaps I should count my proverbial blessings and be happy that all I had to deal with was losing my birthday to a holiday. It could clearly have been much worse.

110 Love Street

Catie Wilkins

I remember being confused as a four-year-old as I sat in assembly at primary school and everyone said the Lord's Prayer. I did as I was told and joined in, saying, "Our Father, who art in heaven." But I thought we were thanking our dads for working hard at their jobs to bring us, their families, home our daily bread, so that we could have Marmite on toast, and jam sandwiches, and other nutritious bread-based snacks. I remember thinking that perhaps I wasn't really eligible to join in anyway, as my dad didn't actually work in heaven—he worked for Tesco. I kept my fears under my hat but felt like a potential fraudster.

My dad, a supremely rational man even when addressing four-year-olds, answered my question "What happens when you die?" logically and truthfully. He replied, "No one really knows, but we have lots of theories. Some people believe in heaven and hell, some people believe in reincarnation, and some people believe that nothing happens." The other four-year-olds were not privy to the open, balanced information that I had, leaving me the only four-year-old to suggest that heaven might not exist. Unlike John Lennon's song "Imagine," this suggestion was not met with delight or praise or musical accolades. The other children just said I was wrong. I became more of an outsider.

I guess I must have continued to feel like an outsider, as when I was five I attempted to send a Christmas card to the Devil. Not to rebel—I was trying to cheer him up. I sent one to God as well, to keep it fair. I wasn't taking sides in their cosmic disagreement.

The card to God (complete with made-up address, 110 Love Street) said, "Well done, you must be very proud." The card to the Devil (who of course lived at 110 Hate Street) said, "Please try to have a good time, in spite of everything." I guess I thought he might be feeling blue or left out on the birthday of his arch-nemesis.

But I think I could relate more to the Devil, and could associate more with his underdog status of everyone hating him. I was chucked out of ballet at the age of four for being disruptive, so I think that the Devil and I both knew what it was like to be excluded from things—he from the eternal paradise for rebelling against the supreme being, I from a ballet class for finding it hilarious to say "no" instead of "yes" when the register was called.

I didn't expect the Devil to write back. Everybody knows he's a bad boy. But God didn't write back either, and he had no excuse. I'd heard the phrases "Ask and you shall be given" and "Seek and ye shall find," but I had scientific evidence that Father Christmas was more communicative than either of them. I'd seen that he'd eaten the mince pies I'd left out for him, but when I'd asked God if I could become a mermaid, my legs had stayed resolutely in place.

However, I decided it was understandable that God was far busier than Father Christmas. After all, while they were both very old and had to keep their long white beards in shape, God had to work 365 days a year (except for Sundays), while Father Christmas only worked for one night, and he also only had to help children, not adults, leaving him more time to stuff his face with mince pies. I guess Father Christmas just had a better union.

I think I partly wanted to become a mermaid because of the biblical story of Noah's Ark—if it happened again, at least I'd be able to swim away. I had always been a bit worried about this story from an animal rights perspective: the other children enjoyed the bit where the animals went in two by two, but I felt sorry for those who hadn't made it onto the ark. For them, it must have been like an animal-based *Titanic*. My one consolation was the fact that all the sea creatures (including dolphins and sea horses) would have survived.

I officially called myself an atheist from the age of ten. I was the only atheist in my class, but the other kids and I did agree on one thing: I wasn't going to heaven. (Though my reasoning was that you couldn't go somewhere that didn't exist.)

I had one ally in our physics teacher (who was an atheist, even though it was a Church of England school). He told us the various things humans have believed about the world, from it being flat to the sun going round the earth, and also told us about the various scientists who had been killed or imprisoned for making new discoveries that went against the doctrine of the church at the time.

He also made a joke that delighted me. Gesturing at the whiteboard, he said, "People used to believe that heaven was up here, earth was in the middle, flat, and hell was down there, below Earth. Which of course we now know can't be true, because hot air rises, and all the people in heaven would have got burned."

This teacher said that science was like a box, and that we could never open its lid. We could, however, investigate in other ways: we could conduct experiments and try to re-create events to get the same results. So we could build an identical box, the same weight and size, and say, "I have discovered what is in the box"; but then, if the first box suddenly turned green but our box didn't, we would have to conclude, "Okay, I was wrong," and start again to try to make our own box go green. In this way science was always learning, changing, and expanding, but admitted to not being absolute.

When I heard that the money from this book was going to go to the HIV charity Terrence Higgins Trust, I was really glad it was going to such a fantastic and worthwhile cause. And it seems appropriate that money raised from a book by atheists is going toward humans helping humans, in both a literal and practical sense.

December is historically a time when humans have a festival to cheer them up because the sun has gone, and Christmas holds the current title. Christmas has done well, to its credit. It's beaten off the competition and is the reigning champion.

There's also a lot to be said for Christmas. The high spirits, good food, and bringing people together are excellent things for humans. Although anyone who says it is the greatest story ever told clearly hasn't read *Watchmen*.

Now that I am an adult, I can look back on the things that used to make me feel confused, alienated, and excluded as an atheist, and take the positives. And in retrospect, sending a Christmas card to the Devil is ironically possibly the most Christian thing you can do—what with all those parables about turning the other cheek.

So my advice to anyone wanting to celebrate an atheist Christmas would be: imagine there's no heaven, then try to have a good time in spite of everything.

Losing My Faith

Simon Le Bon

I love Christmas. I always have, ever since I was a child. Back then, Christmas was all about the Baby Jesus—my parents encouraged belief in him. But even if they hadn't, church and school—which were both Church of England—would have greatly influenced my beliefs.

School was very Christian. At Christmas, we had Nativity plays, but I never got a good role in them. I think I was a sheep. I always believed I was destined for great things there, but I never achieved them.

However, though I was Christian and believed in Jesus, I remember that at school there were these fascinating children who were excused from assembly. They didn't have to attend, and for a long time I thought this was because they were atheists. It was only later that I realized this was because they were Jewish, or Muslim, or Hindu.

I was fascinated by the fact that they were allowed to stay out—I would have loved to. While everybody else was in assembly, you could have wandered around the whole school by yourself without anybody watching you. That was my fantasy—to get up to mischief in the back of the art room.

I had a lot of faith at one time. I was tempted to go to church as a child, because they told me you earned a shilling every week for singing in the choir. I thought, "Mmm, wages!" and became a choirboy.

When you're in a church choir, you actually go to church about five times over Christmas. You go twice on Christmas Eve, and three times on Christmas Day, if you're doing matins, the communion service, and evensong. So that's potentially five professional engagements for a shilling a week over Christmas. The music and the choir was very important to me, and it gave me this feeling of godliness, which I really liked—and I prayed.

But I don't miss that feeling—when it went, it went. It was like somebody pulled the plug out of the bath and the water went down. It didn't feel good while it was going down, but by the time it had gone you'd got used to your body weight, got out of the bath, and got on with something else. That's kind of how it was.

Losing my faith was very gradual. I was confirmed, and I absolutely 100 percent believed in the Christian God. And then, after a while, it started to change. I started losing my faith when I started trying to figure out what God was: "He can't really look like us! This whole thing about how man created God in his own image . . ."

When it came to working out what I really believed in, I realized that, if there is a God, he doesn't have a personality. He certainly doesn't have a set of morals—certainly not human morals, which we impose. And then I started thinking, "Well, what if it's just people trying to personify life? To personify the fact that there is matter, and that there is a universe? If there is a God, that's it. God doesn't have a brain, God doesn't think, God is just existence."

And when you get to that point, you realize that if that's what God is, then there's no such thing.

• • •

For me, the hardest thing about losing my faith was facing the possibility that this life is all there is. One of the foundation stones of all religion is people's fear of death and nonexistence. People will do anything and believe anything if they can think, "You don't really die. There's somebody up there who says you carry on and you go to heaven."

The Buddhists believe in reincarnation, but I tend to think it's rather unlikely that we're going to come back. However, I think there's strength in agnosticism, because you accept that there are things that you cannot know—I cannot know if I've ever existed before this life, and I cannot know if I'm going to exist again. The idea of faith is almost as though, "If I believe it enough, it'll be true." It's a romantic ideal that just doesn't wash with me—I'm too logical.

It's a hard truth, because our instinct is to survive and to continue existing, but I've come to accept that this is it. I'm not scared of not existing. Socrates said that death is unconsciousness, that there's nothing to fear.

I don't want to die, and I'm scared of things that can kill me, so there is a dread of not being around, of not experiencing things, of not seeing the sun rise in the morning, of not knowing what goes on in the world, of not being part of it. But that's normal. There's nothing I can do about it: it's the one great truth, that we all die—you just have to accept it. I hope that when I do die, it'll be at a point when I'm completely ready for it.

I quite like the Atheist Bus Campaign slogan, "There's probably no God." I didn't like it at first—I thought it was too nice. I thought they should have been harder, and wanted them to say, "There's no God, so forget it! You're living in a dream world!" But then it made sense to me, because probability is one of the things I really believe in, in a scientific sense. It's quite healthy to have an open mind.

Religion helps people cope with many things. It helps them deal with death. And I believe in marriage—I doubt the institution of marriage would have existed without religion. To some extent, religion has upheld essential morals and modes of behavior. There are some really important values in all religions.

However, I think human beings go through different stages. As a child, you have someone looking after you. And then you start to break away from that, and eventually you achieve a degree of independence from your parents. Maybe humanity needed a parent and that was the part religion played. Maybe we're at a stage now where we are growing up and ready to achieve a greater degree of independence.

Although it's very tempting to defer responsibility to God, I would like to see humanity taking responsibility for its own actions. There's a certain bravery in standing up and saying, "We are alone, there's no one looking after us." It's a kind of liberation.

Despite having lost my faith, I still celebrate Christmas and I love church music. I go to church to listen to the music. But there's a definite school of thought that says, "If you don't believe it, you can't celebrate it. If you don't believe in God, you can't have Christmas. Sorry—you're excluded!"

To me, it's important that people can believe whatever they like. I'm a liberal; I'm just not religious. If someone else wants to believe in God, they have every right to. I always felt I had the right to believe when I was a Christian.

Most atheists and agnostics feel the same way—we say, "Okay, if you want to believe that, that's fine." Everyone must discover and develop their own beliefs.

Part of me would like there to be a God, because part of me wants there to be a parent looking after me. Someone to say, "Hey, it's okay, it's all under control. No matter how much you mess up, I'm here to save you." That's a very natural feeling, very normal. But on the other hand, I don't think it's enough. I've found I'm more responsible, freer, and more liberated living a life without God. And I love my freedom. I think we all overestimate our freedom. In reality, the freedom to think, to feel, and to experiment is one of the few freedoms we have left.

Hark the Herald Villagers Sing

Zoe Margolis

My first encounter with religion was when I was six years old. At school one day, my teacher told me that I couldn't be in the Christmas Nativity play because I wasn't the "right religion." I remember returning home, crying, devastated that all my friends were going to be having fun in rehearsals, and I would be left alone without their company at break time. And, more importantly—to a six-year-old wannabe actress—I would miss out on the fame and stardom from acting in the play, which was to be performed in front of the entire school. Not to mention not receiving the free sweets used as bribes by the staff for good behavior; I would do anything for a strawberry cream, me.

Brought up in an atheist household, I didn't understand what my teacher meant by "religion": for some reason I thought it suggested I had the lurgy or that something was wrong with me. If I was the "wrong" religion, then surely I could try to become the "right" one and then be part of the school play?

That night, my parents patiently tried to explain the concept of "God" to me. I must admit, being the snotty-nosed brat that I was, who absorbed books like oxygen, I was slightly impressed by their bringing out a copy of the Bible to show me, while they attempted to condense a few thousand years of religious doctrine into a child-friendly atheist version. But even then I was cynical: I'd learned early on that the tooth fairy was pretend, and I'd recently discovered that Santa Claus was purely fictional (and was pretty devastated by that), so why should I believe in this God bloke? It's not like I'd ever seen any evidence of him—and he'd certainly never left me any coins under my pillow or filled the stocking at the end of my bed with presents. What had God ever done for me besides prevent me from getting a starring role in the Christmas play? Even then, I knew I didn't like him. And that whole burning bush thing scared me a bit, if I'm honest.

The following day, my mom grabbed me by the arm, stormed into the school, and had a huge argument with my teacher; I remember lots of heated words being exchanged. Back then, I just thought my mom was defending her prima donna daughter; it was only as an adult that I learned she had accused the teacher of discrimination. I now understand and appreciate the importance of my mom sticking up for her atheist beliefs and the right of her child not to be subject to prejudice because of them.

The teacher finally caved in to my mom's persuasiveness and agreed to let me have a part in the play. I was joyous with happiness. Now I would have fame! Glory! Attention! Me, as Mary! (Whoever she was; I didn't care. That was the lead role, and I wanted it.) Or as an angel! (Again, not sure what or who they were, but if they got to flutter around in a white tutu, I was more than game.) I was so excited: I could almost see my name in lights. Almost.

I bounced around the rest of the day and, like the precocious diva I was, looked forward to my costume fitting. And when it came, I lined up with all my friends and waited for my name to be called as the roles were divvied up in alphabetical order. (This has been the bane of my life, given my name begins with a Z. Last in line for everything.)

"Ashling!" my teacher called, and my friend was given the role of Mary.

Damn. Lost the lead role. Oh well, I will still be a pretty angel!

"Cathy!" the teacher said, and proceeded to make my best friend an angel.

I grew ever more excited, though: I couldn't wait to try on the tutu!

"Fiona!" the teacher barked, and my friend went off to get her tutu fitted.

It would be me soon! *Tutu, here I come!*

"Helena!" shouted the teacher, and yet another friend was sent to the angel queue.

This went on for a while, until there were a dozen angels, as well as a few wise men, and only a couple of us left standing in the queue.

I think I knew at that point that my hopes of having a starring role were about to be severely dashed. But—ever the (non-eternal, reincarnation-cynical) optimist—I thought that perhaps I would be made a special angel: a lead angel who was in charge of all the other angels and who got to boss them around and stuff. Maybe I could wear a black tutu instead, like in *Swan Lake*?

My name was finally called: I was at the back of the line, there were few costumes left, I was the last pupil to be given a role.

"You're going to be a villager in the choir," my teacher informed me.

I stared at her, gobsmacked.

"Tell your mom that you will need to bring a scarf, gloves, and hat with you to wear to all the rehearsals."

Oh, great, I don't even get a costume. My dreams of stardom vanished in a second.

"And," my teacher continued, "you get to hold this lantern. Isn't it nice?!"

I think, even back then, I knew she was being sarcastic. Bitch.

My teacher handed me a long wooden stick with a pretend lantern dangling on one end.

And it was at this point that I had a stroke of genius: a way for me to decline this minor, irrelevant role, and be promoted into a proper acting part.

"I can't hold that," I said.

"Why not?"

"I'm allergic to wood."

I don't know if she was more surprised by the absurdity of what I had said or by the fact that it had been said by a smart-alecky, upstart six-year-old, but whichever it was, she wasn't pleased. She wrote a huffy note, which I gave to my mother later, that said I had been offered a role but was now making up lies to get out of it.

My mom sat me down that night and asked me what I wanted to do (while sniggering about my wood allergy comment, I should add). My only options, it seemed, were either not be in the school play at all or accept the role of an extra and perform in the choir. With all my friends already practicing their lines, and not wanting to be left out, I chose the latter.

Photographs taken of the play, when it was performed some weeks later, just before Christmas time, show a very cheery Mary and Joseph, some happy wise men, many elegant and joyous angels, and, standing in the back of the villagers' choir, one extremely pissed-off, scowling six-year-old, holding her lantern askew. Let's just say I was not at all happy.

Years later, when I look back on that event, it seems clear to me that that was the defining moment when I realized I could not believe in God. Sure, as an adult, surrounded by science and reason, it's obvious to me that God doesn't exist. But, as a starry-eyed six-year-old, my disbelief in religion came down to three simple facts:

1. I never got to eat a strawberry cream, because being last in line all the time meant everyone else had already nabbed them. (God can't be that cruel, surely.)
2. I did not achieve international stardom from my role as a villager. (God can't be that mean, surely.)
3. Anyone who would allow a child to be forced to sing "Hark! The Herald Angels Sing!" is a sadist, not a deity. (I am assuming God is not into S&M.)

Although I suppose it could be argued that God might exist, for the world at large was prevented from being exposed to my performing at a professional level. Given my singing voice, that really is something to rejoice and say "Hallelujah!" over.

The Christmas Miracle Event:

A Story

Evan Mandery

Nathan Townsend's father had only ever given him two pieces of advice: One, don't fight a two-front land war in Europe. Two, don't get drunk at the office Christmas party. This wasn't much of a parental legacy, but it seemed like good advice and, up until that evening, he had faithfully followed both of Herb Townsend's maxims, the first with little or no inconvenience, though the second with some, given Nathan's fondness for eggnog, particularly eggnog with whisky or brandy or rum, each of which was present in abundance at the party, and which, Nathan learned that night, combined to surprising and substantial effect.

Now, the holiday gathering at the Ludwig Andreas Feuerbach School of Ethics and Moral Culture wasn't specifically a Christmas party. That really wouldn't fly with the PTA. Technically, it was a celebration of Juleaftensdag, the eve of the pagan winter festival, Yule. The party-planning committee aimed to be broadly inclusive and incorporated elements of many different Germanic heathen cultures. Dried straw was laid across the floor of the gymnasium following the Estonian custom. Each of the first and second graders left a shoe in the window, in the hope they might be visited by one of the thirteen Yule Lads, who imperfectly filled the role of Santa in the Icelandic tradition. Some of the Yule Lads leave modest gifts like potatoes or apples. One helpful fellow, called the Pot Scraper, cleans the kitchen pans.

The food was similarly multicultural. They had beetroot salad from Finland, roast goose with red cabbage representing Denmark, and *sill* (pickled herring) from Sweden. The pièce de résistance was a traditional Icelandic food from the Westfjords: fermented *skata* (stingray) with melted tallow and boiled potatoes. The spread was authentic but grim. Nathan tried the *skata* and thought it tasted like spoiled bologna. But the liquor was good, and thus Herb Townsend's second adage should have applied, particularly given that Nathan ate nothing other than the sliver of stingray and lost count of the glasses of nog.

The trouble began when Nathan got roped into a conversation with Potter Everson. Nathan hated Potter Everson. Potter taught the advanced placement course in the history of cynicism and three years running had been voted the senior class' most coveted superlative, Least Likely to Inspire. He paraded around the school like a big man on campus, wearing suede moccasins, Madras shorts, and cardigan sweaters, a combination he described as "postmodern hip." At any normal school, the students would have mercilessly teased Potter Everson into an insane asylum, but the students at the Feuerbach School were, by persistent training, tolerant of almost everything (with one notable exception). They embraced Potter Everson. Nathan, however, avoided him at all costs.

But thanks to the rum or the brandy or the whisky—he couldn't be sure—Nathan's guard was down, and when Joe Kafka, a grizzled veteran of the science department, grabbed him by the arm and said, "You've got to hear this one," Nathan hardly had time to protest. Before he knew it, he was standing with Joe in a large circle that included, among others, Ellen Nordberg, the principal's secretary, and Flip Anderson, the custodian who regulated the pool's chlorine content, listening to Potter Everson tell a hilarious story.

"So I'm standing on the corner of Seventy-second Street and Broadway waiting for the bus, and the Lubavitchers are out in force. The Mitzvah-mobile is parked on the corner and they're scouring the intersection, in full regalia, sloughing off menorahs on unsuspecting pedestrians. They approach the bus line and ask, one person after another, 'Are you Jewish? Are you Jewish?' Everyone ignores them until they come to this man at the end of the line. He's wearing a derby and a gray raincoat and looking generally meek and vulnerable. 'Are you Jewish?' they ask. He looks at his shoes and sheepishly says, 'I'm an agnostic.'"

The group howled with laughter. In any other environment the story would not have been regarded as especially funny. It wouldn't even have qualified as a joke. But at the School of Moral and Ethical Culture, agnostics were regarded with the same derision reserved in the general population for the Polish, hillbillies, and congressmen. As with these comically disfavored minorities, agnostic jokes had become something of an art form. Thus the favorable reaction.

Potter went on. "'In that case,' the Lubavitcher said, 'you might want to take a menorah—just to be safe.'" Within the circle, chortles and smirks were suppressed, as the faculty and staff eagerly anticipated the punch line.

"So what does the guy in the gray raincoat do?" The group was ready to burst. Wait for it.

"He takes the menorah!"

Hereupon followed even more voluble howls of laughter, shortness of breath, and general glee. Ellen Nordberg grabbed her stomach to keep from keeling over. Flip Anderson wiped tears from his eyes. Doris Keeling, the third-grade teacher, suffered a paroxysm. The wave of euphoria infected everyone except for Nathan, who did not find the story amusing at all. To the contrary, he found it decidedly annoying.

In retrospect, Nathan would find it difficult to explain why he had such a negative reaction to the joke. He had heard agnostics made fun of many times. While, for a variety of reasons, he didn't find the jokes particularly funny, he didn't regard them as offensive, since religion, unlike race or ethnicity, was a matter of personal choice. Thus, within the precise ethical code of the school, the subject was fair game. But a negative reaction he had all the same. Whereas everyone else was in hysterics, Nathan groused and frowned and moped. Without thinking, he muttered, "I'm an agnostic."

Ellen Nordberg spit out her raspberry seltzer. She thought it was part of the joke.

Joe Kafka hit him on the back and said, "That's a good one, Nate."

But Nathan said, "No, I'm serious," with a look that showed he really was. "I'm an agnostic," he repeated.

The room rapidly deflated. Nathan Townsend was both lucky and unlucky in this moment. At another school, they would not have cared that he was an agnostic. So that was unlucky. But at another school where agnostics were scorned, they might have openly derided him. On this count Nate was fortunate. In the rigorously precise ethic of the Feuerbach School, it was decidedly unacceptable to mock someone to his face, no matter what the offense. Thus no one dared ridicule Nathan openly.

But leprosy would have been a more popular admission. At the Feuerbach School, it went without saying that everyone was expected to be an atheist, or more precisely a secular humanist. Secular humanism is a value system that embraces reason and justice and rejects religion as a basis for moral decision making. This repudiation of dogma was utterly essential to the culture of the school. The following quotation was emblazoned above the main door, through which each student and faculty member walked every day:

Religion is all bunk.
—Thomas Alva Edison

So fervently committed to this outlook on life was the leadership of the Feuerbach School that it had been trying for years to have the American government recognize secular humanism as a religion. To the indoctrinated observer, this might appear to be a contradiction, but to the Feuerbach School it was a matter of high principle. Also, the designation carried with it certain tax advantages. Inevitably, the debate had devolved into litigation, which was not going well for the school. The trial judge, and the members of a unanimous appellate court panel, had all attached great weight to the Edison quote. In the view of the school's attorneys, this was too literal a reading of the quote above the door. For example, the following colloquy occurred during the argument before the Second Circuit Court of Appeals sitting en banc:

Judge Hiram Fernandez: Counsel, how can you argue that the Feuerbach School has a religious mission, and hence is entitled to IRS Code Section 501(c)(3) status, when the sign above your door says "Religion is all bunk?"

William Daley, Esq. (of Daley, Daley, Daley & Dealey): Your Honor, with all due respect, I submit you are reading the inscription too literally.

Judge Hiram Fernandez: Oh? How should I read it?

William Daley, Esq.: The critical point is that my clients are zealously dedicated to their absence of faith-based conviction. One might reasonably say that they are . . . religious about it.

Judge Hiram Fernandez: I see. Very clever.

In truth, Hiram Fernandez was not impressed at all and, as every lawyer knows, as goes Hiram Fernandez, so goes the Second Circuit.

The school lost the appeal 9–0. Bill Daley blustered to the press about an appeal to the Supreme Court, but no one at Feuerbach held out much hope, and the case was something of a sore spot at the school.

"Don't you believe in the scientific method?" Joe Kafka asked incredulously.

"I certainly believe in the fact of it," Nathan said. "And I believe it has produced many socially useful results."

"Then why are you willing to make a leap of faith and say that God may exist?"

"Let me ask *you* this," Nathan replied. "Applying scientific evidence, what evidence is there that God does not exist?"

"Only my experience," Joe said.

"Is it not then a leap of faith to say that he does not exist?"

Joe Kafka had no answer. Silently he took a step away from Nathan. So did Potter Everson and Ellen Nordberg. Even Doris Keeling retreated, and she could tolerate almost anything: her husband had been having an open affair with a goat for thirty years. One by one, the faculty and staff filtered away until only Flip Anderson remained.

Quietly Nathan asked, "Flip, do you really believe there's no chance that God exists?"

"Well, he sure don't keep the chlorine levels straight," Flip replied. Then he walked away, leaving Nathan alone.

At Hi Life, on Eighty-third and Amsterdam, Nathan related the story to his friend, Lou Pinto, an eastern gray kangaroo whom Nathan had met several years earlier at a Bikram yoga class. Nathan hadn't stuck with it—the moist heat aggravated his sinuses—but Lou had, which was ironic in a way, because he had a bit of a temper and, generally speaking, didn't seem the yogic type. Nevertheless, he was now enviably flexible.

Lou lived all the way up in Morningside Heights, but he was always happy to go out for a drink. He didn't sleep much, and besides, it was Christmas Eve. The Hi Life patrons were dudded up in festive reds and greens with floppy Santa Claus hats hanging from their heads, drinks in their hands, and cheeks aglow. The room was abuzz, abounded in good holiday cheer, which had affected everyone—everyone, that is, except for Nathan Townsend. Lou Pinto noticed as soon as he hopped in.

"Who died?" he asked. "You look like someone killed Santa Claus."

Nathan told him what happened.

"That's too bad," Lou said with obvious sincerity. Lou was a good and patient listener. "Why do you think you said it?" he asked. "Was it the nog?"

"Maybe," Nathan said, "But I really think it was that Potter Everson. Something about him always throws me off."

Lou nodded. "What do you think is going to happen to you?"

"Nothing, I suppose."

"You have tenure, after all."

"True."

"So why are you beating yourself up about it?"

Nathan sighed. "I'm jealous," he said. "I look at my colleagues at the school with envy. They have conviction about things, about what's right and wrong, and the ultimate direction and meaning of life. It gives them a sense of purpose and certainty. Perhaps it's smug on their part. I don't know. Call it what you will, but it seems like a happier life they have."

"You can have that life too," Lou said.

Nathan's face revealed his inner turmoil. "It isn't as easy as that."

"Why not, brother?"

"I have doubts! I have such doubts!" Nathan said. "I look around me and what I see fills me with awe. In this city are animals of every kind, living together. I read about mountains that touch the sky and volcanoes miles under the sea. Soaring birds, glorious plants, fish of every size, shape, and color imaginable. Such wondrous life! Could it truly be all random? Perhaps, but can anyone be sure?" He look at up from his drink and faced Lou. "Don't you ever have doubt?"

This was a bad question. Lou's father had been grossly abusive. If he failed to clean his room or talked out of turn, his father would box him, often in public. Years of therapy and tantric meditation had helped Lou release the anger, but he had no uncertainty about the absence of God.

"No," Lou said.

"How can it be that easy for you?"

"It's just a matter of faith," Lou said.

Nathan nodded. "If only I could make the leap," he said.

Lou smiled. "We've both probably had enough to drink," he said. "How about I walk you home?"

"It's all the way on the East Side, completely out of your way."

"What are friends for?"

Lou was a good one. He even pulled his wallet from his pouch and picked up the tab.

The Central Park Reservoir can be spectacular on a winter evening. In the cold crisp air, the lights of the New York skyline reflect off the water. It is a sight without parallel in the great cities of the world, and that evening was as fine as any there had ever been. Lou Pinto appreciated its beauty. He had jogged around this track countless times, often late at night, but he had never seen anything like this. He hoped the nighttime splendor would cheer Nathan, but it had no effect. His friend stared at the ground as he trudged along, moping.

Lou felt his friend's pain. He did not specifically understand the angst of a crisis of faith. Lou had never questioned the nonexistence of God. For him this had always been axiomatic. But he knew what it was like to be different from those around you, to be an outsider. This anguish he understood all too well. He placed a gentle arm around the shoulder of his friend and tried to absorb some of the hurt. Sadly, it had no effect.

As they neared the South Gatehouse, a fog began to roll in. At first it did not seem unusual, other than the fact of it on such a luminous night, but it progressed with preternatural dispatch and had to it an otherworldly density and odor. In the course of taking no more than twenty steps, Nathan and Lou had moved from crystal-clear Christmas air to a dense, miasmic thicket. Instinctively, they turned to retrace their steps, but the brume had closed in on them from behind too, and presently there was no escape. They took a few clumsy steps in the direction they believed to be forward. Lou, normally sure of foot, stumbled, and then they stopped in their tracks.

Just as he spoke, an ill wind blew through. It sent a chill up Nathan's spine and activated the dull ache in Lou's trick knee, which he had injured twenty years earlier running the Tel Aviv Marathon. Lou sensed trouble afoot. The joint only bothered him when something bad was about to happen. The last time, he'd got home from work to find that a wallaby he was dating had absconded with his collection of Richie Rich comic books.

When the gust died down, the air immediately before them had cleared a bit, just enough so that Nathan and Lou could see standing before them the apparition of a figure from the past. With muttonchops, long coat, and vest, he could have been any nineteenth-century English gentleman out for a postprandial constitutional. But he was unmistakably the famous English biologist—staunch advocate of Darwin, notorious verbal sparring partner of Samuel Wilberforce, and coiner of the term "agnostic."

"Thomas Henry Huxley," Nathan and Lou said softly, in unison.

"Which one of you is Nathan Townsend?" the specter asked.

"Who wants to know?" Lou asked protectively. It seemed odd that this apparently supernatural creature didn't know whether Nathan was a human being or a kangaroo.

"I have been sent with an important message for Nathan Townsend."

"How do we know you're authentic?" Lou repeated. The ghost's voice had the appropriate resonance and vibrato, but Lou wanted hard proof.

"When he was seven years old he had to have an impacted marble surgically removed from his right nostril."

"Anyone could know that. It's public record."

"It was, specifically, a blue marble from the game Mouse Trap. He was frustrated because the swinging boot never functioned properly."

Lou looked at Nathan.

"It's true," Nathan said. "He's for real."

Lou looked back at the phantasm. "What is it, then? What do you have to say?"

"I have been told to tell you this conclusively once and for all." Here the ghost of Thomas Henry Huxley paused for dramatic effect. "There is no God!" The proclamation resonated through the red maples and pin oaks of the Ramble. In the distance, an owl hooted.

"How do you know this?" Nathan asked.

"I have been told so by an omniscient, all-knowing being whose credibility I can personally verify."

"Then that would be God," said Lou.

"No," said Huxley. "He sees and knows all, but his power has limits."

"Such as?"

"He is very poor at golf. He has taken lessons for thousands of years, but he still slices. He also has great difficulty getting the chlorine right in his hot tub."

"I've heard that can be difficult," Nathan said.

"He knows of your anguish, Nathan Townsend. He wants your mind to be at ease."

"Thank you," said Nathan.

Huxley turned. "And Lou Pinto, he told me to tell you that your father loved you very much."

Tears formed in Lou's eyes. It was the greatest gift anyone could have given him. "Thank you," he said. "Thank you so much."

But the ghost of Thomas Henry Huxley was already gone.

When the fog lifted, Nathan felt as if an enormous weight had also been lifted from his own shoulders. It didn't seem it should be so easy to secure peace of mind, but Huxley had given it to him. Now he had certainty and conviction, and this in turn made him buoyant. Nathan resumed their walk home with a spring in his step, and an unfamiliar sense of optimism.

"You look like a new man," Lou said, who felt a great sense of relief himself. "It's a Christmas miracle."

Nathan corrected him. "A miracle event," he said. "A Christmas miracle event."

Lou smiled, and together they walked into the holiday night.

SCIENCE

I do not believe in a personal God and have never denied this but have expressed it clearly. If there is something within me that can be called religious then it is the unbounded admiration for the structure of the world so far as our science can reveal it.

—Albert Einstein

The Sound of Christmas

SIMON SINGH

While Christians celebrate the birth of Jesus at Christmas, atheists may wonder if there is another birth they might be able to commemorate. One possibility is to give thanks for the arrival of Isaac Newton, who was born on Christmas Day 1642 according to the Julian calendar, which was still in use in England at the time. Another possibility, and probably my preference, is to use Christmas Day as an excuse to celebrate the biggest birth of all, namely, the creation of the entire universe.

For tens of thousands of years, humans have stared up into the heavens and wondered about the origin of the universe. Up until now every culture, society, and religion has had nothing else to turn to except its creation myths, fables, or religious scriptures. Today, by contrast, we have the extraordinary privilege of being the first generation of our species to have access to a scientific theory of the universe that explains its origin and evolution. The Big Bang model is elegant, magnificent, rational, and (most importantly of all) verifiable. It explains how roughly 13.7 billion years ago matter exploded into being and was blown out into an expanding universe. Over time this matter gradually coalesced and evolved into the galaxies, stars, and planets we see today.

Before explaining how you might celebrate the birth of the universe, let me quickly explain why we are convinced that there was a Big Bang. First of all, telescope observations made back in the 1920s seemed to show that all the distant galaxies in the universe were redder than they should have been. Red light has a longer wavelength than all the other colors, so it was as if the light from the galaxies was being stretched. One way to explain this stretching of galactic light (otherwise known as the "red shift") was to assume that space itself was expanding. Expanding space is a bizarre concept, but it is exactly how we would expect space to behave in the aftermath of a Big Bang explosion.

However, this single piece of evidence was not enough to convince the scientific establishment that the Big Bang had really happened, particularly as the observations were open to interpretation. For example, the Bulgarian-born astrophysicist Fritz Zwicky pointed out the redness of the galaxies was merely an illusion caused by the scattering of light by dust and gas as it passed through the cosmos.

By the way, as well as being a critic of the Big Bang and the data that seemed to support it, Zwicky was also responsible for inventing a beautiful insult. If a colleague annoyed him, Zwicky would scream out, "Spherical bastard." Just as a sphere looks the same from every direction, a spherical bastard was someone who was a bastard whatever way you looked at him.

A second pillar was needed to support the Big Bang model, and this time the crucial evidence relied on measuring the ingredients of the universe, most importantly hydrogen and helium. These are smallest atoms in the periodic table and the most common in the universe, accounting for 74 percent and 24 percent of all atoms, respectively. Crucially, the only way to create such large amounts of hydrogen and helium is in the wake of the Big Bang. In particular, the pressure, density, and temperature of the early universe would have cooked exactly the right amount of hydrogen and fused it into exactly the right amount of helium. In other words, the Big Bang is the best (and probably the only) way to explain the abundances of these light elements.

Nearly all the other elements were made later in collapsing stars. These stars provided the perfect environment for the nuclear reactions that give rise to the heavier elements that are essential for life. Marcus Chown, author of *The Magic Furnace*, highlighted the startling significance of stellar alchemy: "In order that we might live, stars in their billions, tens of billions, hundreds of billions even, have died. The iron in our blood, the calcium in our bones, the oxygen that fills our lungs each time we take a breath—all were cooked in the furnaces of the stars which expired long before the Earth was born."

Because we are made from the debris of nuclear reactions that took place in exploding stars, the romantics among you might like to think of yourselves as being composed of stardust. On the other hand, cynics might prefer to think of yourselves as nuclear waste.

The third, and even sturdier, pillar to support the Big Bang model is the afterglow that should have followed a creation event, which can still be seen today. The theory behind the Big Bang suggests that intense shortwave radiation was released just a few hundred thousand years after the initial expansion. This radiation would have been stretched as the universe expanded, meaning that it would exist today in the form of microwave radiation. These microwaves from the Big Bang should still exist in all parts of the universe at all times and are therefore an excellent make-or-break test for whether or not the universe did start 13.7 billion years ago.

Although the Big Bang microwaves were predicted in 1948, they were soon forgotten because astronomers did not have any technology that was sensitive enough to detect microwaves from outer space. However, in 1964 two American radio astronomers discovered them in an episode of pure serendipity. (Serendipity is the art of making fortunate discoveries by accident, or as one anonymous male scientist put it: "Serendipity means looking for a needle in a haystack and finding the farmer's daughter.")

Arno Penzias and Robert Wilson were using something called a radio telescope to study galaxies. A radio telescope is a large dish or cone that detects radio waves instead of visible light waves. Annoyingly, the astronomers noticed that they were picking up unexpected radio waves coming all the time from all directions. Initially, they thought the signal might be an error caused by a component within the telescope, so they began to check every single element of the equipment. They searched for dodgy contacts, sloppy wiring, faulty electronics, misalignments in the cone, and so on.

When they climbed inside the cone they discovered a pair of nesting pigeons that had deposited a "white dielectric material." Thinking that this pigeon poo was somehow causing the spurious signal, they trapped the birds, placed them in a delivery van, and had them released thirty miles away. The astronomers then scrubbed and polished the cone, but the pigeons obeyed their homing instinct, flew back to the telescope, and started depositing white dielectric material all over again. When I met Arno Penzias in 2003, he described to me what happened when he recaptured the pigeons: "There was a pigeon fancier who was willing to strangle them for us, but I figured the most humane thing was just to open the cage and shoot them."

Of course, even without the pigeons and their pigeon poo, the microwaves still kept coming, and after several weeks Penzias and Wilson eventually realized that they had discovered the leftover radiation from the Big Bang. This was one of the most sensationally serendipitous discoveries in the history of science, and a decade later the lucky duo were rewarded with the Nobel Prize for essentially proving that the Big Bang had really happened.

Some people sneer at the accidental nature of this discovery and question whether it deserved the Nobel Prize. Such folk would do well to remember the words of Winston Churchill: "Men occasionally stumble over the truth, but most of them pick themselves up and hurry off as if nothing ever happened." Indeed, it seems likely that other astronomers probably detected the microwave radiation from the Big Bang before 1964, but it was so faint that they ignored it and carried on regardless.

In fact, you have probably witnessed this Big Bang radiation yourself without realizing it, because most old radios are capable of picking up microwaves. And because a radio can act as a very, very, very primitive radio telescope, I suggest that you use one as the focus for your Christmas celebration of the birth of the universe. Here's what you need to do.

At some point over the Christmas period switch on an analog radio and retune it so that you are not on any station. Instead of "Jingle Bells" or "Away in a Manger," all you should be able to hear is white noise. This gentle, calming hiss is the audible output caused by all sorts of random electromagnetic waves being picked up by your radio aerial. You cannot single them out, but rest assured that about 1 percent or 2 percent of these waves are due to microwaves from the Big Bang. In other words, your humble radio is capable of detecting energy waves that were created over 13 billion years ago.

While everyone else is pulling crackers or arguing over the last chocolate orange segment, you can simply close your eyes and listen to the sound of the universe. You are experiencing the echo of the Big Bang, a relic of creation, the most ancient fossil in the universe.

The Great Bus Mystery

Richard Dawkins

I was hoofing it down Regent Street, admiring the Christmas decorations, when I saw the bus, one of those bendy buses that mayors keep threatening with the old heave-ho. As it drove by, I looked up and got the message square in the monocle. You could have knocked me down with the proverbial. Another of the blighters nearly did knock me down as I set a course for the Dregs Club, where it was my purpose to inhale a festive snifter, and I saw the same thing on the side. There are some pretty deep thinkers to be found at the Dregs, as my regular readers know, but none of them could make a dent on the vexed question of the buses when I bowled it their way. Not even Swotty Postlethwaite, the club's tame intellectual. So I decided to put my trust in a higher power.

"Jarvis," I sang out as I latchkeyed self into the old headquarters, shedding hat and stick on my way through the hall to consult the oracle. "I say, Jarvis, what about these buses?"

"Sir?"

"You know, Jarvis, the buses, the 'What is this that roareth thus?' brigade, the bendy buses, the conveyances with the kink amidships. What's going on, Jarvis? What price the bendy bus campaign?"

"Well, sir, I understand that, while flexibility is often considered a virtue, these particular omnibuses have not given uniform satisfaction. Mayor Johnson—"

"Never mind Mayor Johnson, Jarvis. Consign Boris to the back burner and bend the bean to the buses. I'm not referring to their bendiness per se, if that is the right expression."

"Perfectly correct, sir. The Latin phrase might be literally construed—"

"That'll do for the Latin phrase, Jarvis. Never mind their bendiness. Fix the attention on the slogan on the side. The orange-and-pink apparition that flashes by before you have a chance to read it properly. Something like 'There's no bally God, so put a sock in it and have a gargle with the lads.' That was the gist of it, anyway, although I may have foozled the fine print."

"Oh, yes, sir, I am familiar with the admonition: 'There's probably no God. Now stop worrying and enjoy your life.'"

"That's the baby, Jarvis. Probably no God. What's it all about? Isn't there a God, Jarvis?"

"Well, sir, some would say it depends upon what you mean. All things that follow from the absolute nature of any attribute of God must always exist and be infinite, or, in other words, are eternal and infinite through the said attribute. Spinoza."

"Thank you, Jarvis, I don't mind if I do. Not one I've heard of, but anything from your shaker, Jarvis, always hits the spot and reaches the parts other cocktails can't. I'll have a large Spinoza, shaken, not stirred."

"No, sir, my allusion was to the philosopher Spinoza, the father of pantheism, although some prefer to speak of panentheism."

"Oh, that Spinoza. Yes, I remember he was a friend of yours. Seen much of him lately?"

"No, sir, I was not present in the seventeenth century. Spinoza was a great favorite of Einstein, sir."

"Einstein, Jarvis? You mean the one with the hair and no socks?"

"Yes, sir, arguably the greatest physicist of all time."

"Well, Jarvis, you can't do better than that. Did Einstein believe in God?"

"Not in the conventional sense of a personal God, sir. He was most emphatic on the point. Einstein believed in Spinoza's God, who reveals himself in the orderly harmony of what exists, not in a God who concerns himself with fates and actions of human beings."

"Gosh, Jarvis, bit of a googly there, but I think I get your drift. God's just another word for the great outdoors, so we're wasting our time lobbing prayers and worship in his general direction, what?"

"Precisely, sir."

"If, indeed, he has a general direction," I added moodily, for I can spot a deep paradox as well as the next man—ask anyone at the Dregs. "But Jarvis," I resumed, struck by a disturbing thought, "does this mean I was also wasting my time when I won that prize for scripture knowledge at school? The one and only time I elicited so much as a murmur of praise from that prince of stinkers, the Reverend Aubrey Upcock? The high spot of my academic career, and it turns out to have been a dud, a washout, scrapped at the starting gate?"

"Not entirely, sir. Parts of holy writ have great poetic merit, especially in the English translation known as the King James, or Authorized, Version of 1611. The cadences of the Book of Ecclesiastes and some of the prophets have seldom been surpassed, sir."

"You've said a mouthful there, Jarvis. Vanity of vanities, saith the preacher. Who was the preacher, by the way, Jarvis?"

"That is not known, sir, but informed opinion agrees that he was wise. 'Rejoice, O young man, in thy youth; and let thy heart cheer thee in the days of thy youth.' He also evinced a haunting melancholy, sir. 'When the grasshopper shall be a burden, and desire shall fail: because man goeth to his long home, and the mourners go about the streets.' The New Testament too, sir, is not without its admirers.

'For God so loved the world that he gave his only begotten Son . . .'"

"Funny you should mention that, Jarvis. The passage was the very one I raised with the Reverend Aubrey, and it provoked a goodish bit of throat clearing and shuffling of the trotters."

"Indeed, sir. What was the precise nature of the late headmaster's discomfort?"

"All that stuff about dying for our sins, redemption, and atonement, Jarvis. All that 'and with his stripes we are healed' carry-on. Being, in a modest way, no stranger to stripes administered by old Upcock, I put it to him straight: 'When I've performed some misdemeanor'—or malfeasance, Jarvis?"

"Either might be preferred, sir, depending on the gravity of the offense."

"So, as I was saying, when I was caught perpetrating some malfeasance or misdemeanor, I expected the swift retribution to land fairly and squarely on the Woofter trouser seat, not some other poor sap's innocent derrière, if you get my meaning, Jarvis?"

"Certainly, sir. The principle of the scapegoat has always been of dubious ethical and jurisprudential validity. Modern penal theory casts doubt on the very idea of retribution, even where it is the malefactor himself who is punished. It is correspondingly harder to justify vicarious punishment of an innocent substitute. I am pleased to hear that you received proper chastisement, sir."

"Quite, Jarvis."

"I am so sorry, sir, I did not intend—"

"Enough, Jarvis. This is not dudgeon. Umbrage has not been taken. We Woofters know when to move swiftly on. There's more, Jarvis. I hadn't finished my train of thought. Where was I?"

"Your disquisition had just touched upon the injustice of vicarious punishment, sir."

"Yes, Jarvis, you put it very well. Injustice is right. Injustice hits the coconut with a crack that resounds around the shires. And it gets worse. Now, follow me like a puma here, Jarvis. Jesus was God, am I right?"

"According to the trinitarian doctrine promulgated by the early Church fathers, sir, Jesus was the second person of the triune God."

"Just as I thought, Jarvis. So God—the same God who made the world and was kitted out with enough nous to dive in and leave Einstein gasping at the shallow end, God the all-powerful and all-knowing Creator of everything that opens and shuts, this paragon above the collarbone, this fount of wisdom and power—couldn't think of a better way to forgive our sins than to turn himself over to the gendarmerie and have himself served up on toast. Jarvis, answer me this. If God wanted to forgive us, why didn't he just forgive us? Why the torture, Jarvis? Whence the whips and scorpions, the nails and the agony? Why not just forgive us? Try that on your Victrola, Jarvis."

"Really, sir, you surpass yourself. That is most eloquently put. And if I might take the liberty, sir, you could even have gone further. According to many highly esteemed passages of traditional theological writing, the primary sin for which Jesus was atoning was the original sin of Adam."

"Dash it, Jarvis, you're right. I remember making the point with some vim and élan. In fact, I rather think that may have been what tipped the scales in my favor and handed me the jackpot in that scripture knowledge fixture. But do go on, Jarvis, you interest me strangely. What was Adam's sin? Something pretty fruity, I imagine. Something calculated to shake hell's foundations?"

"Tradition has it that he was apprehended eating an apple, sir."

"Scrumping, Jarvis? That was it? That was the sin that Jesus had to redeem—or atone according to choice? I've heard of an eye for an eye and a tooth for a tooth, but a crucifixion for a scrumping? Jarvis, you've been at the cooking sherry. You are not serious, of course?"

"Genesis does not specify the precise species of the purloined comestible, sir, but tradition has long held it to have been an apple. The point is academic, however, since modern science tells us that Adam did not in fact exist, and therefore was presumably in no position to sin."

"Jarvis, this takes the chocolate digestive, not to say the mottled oyster. It was bad enough that Jesus was tortured to atone for the sins of lots of other fellows. It got worse when you told me it was only one other fellow. It got worse still when that one fellow's sin turned out to be nothing worse than half-inching a D'Arcy Spiçe. And now you tell me the blighter never existed in the first place. Jarvis, I am not known for my size in hats, but even I can see that this is completely doolally."

"I would not have ventured to use the epithet myself, sir, but there is much in what you say. Perhaps in mitigation I should mention that modern theologians regard the story of Adam and his sin as symbolic rather than literal."

"Symbolic, Jarvis? Symbolic? But the whips weren't symbolic. The nails in the cross weren't symbolic. If, Jarvis, when I was bending over that chair in the Reverend Aubrey's study, I had protested that my misdemeanor, or malfeasance if you prefer, had been merely symbolic, what do you think he would have said?"

"I can readily imagine that a pedagogue of his experience would have treated such a defensive plea with a generous measure of skepticism, sir."

"Indeed you are right, Jarvis. Upcock was a tough bimbo. I can still feel the twinges in damp weather. But perhaps I didn't quite skewer the point, or nub, in re the symbolism?"

"Well, sir, some might consider you a trifle hasty in your judgment. A theologian would probably aver that Adam's symbolic sin was not so very negligible, since what it symbolized was all the sins of mankind, including those yet to be committed."

"Jarvis, this is pure applesauce. 'Yet to be committed'? Let me ask you to cast your mind back yet again, Jarvis, to that doom-laden scene in the beak's study. Suppose I had said, from my vantage point doubled up over the armchair, 'Headmaster, when you have administered the statutory six of the juiciest, may I respectfully request another six in consideration of all the other misdemeanors, or peccadilloes, which I may or may not decide to commit at any time into the indefinite future? Oh, and make that all future misdemeanors committed not just by me but by any of my pals.' Jarvis, it doesn't add up. It doesn't float the boat or ring the bell."

"I hope you will not take it as a liberty, sir, if I say that I am inclined to agree with you. And now, if you will excuse me, sir, I would like to resume decorating the room with holly and mistletoe, in preparation for the annual Yuletide festivities."

"Decorate if you insist, Jarvis, but I must say I hardly see the point anymore. I expect the next thing you'll tell me is that Jesus wasn't really born in Bethlehem, and there never was a stable or shepherds or wise men following a star in the East."

"Oh, no, sir. Informed scholars from the nineteenth century onward have dismissed those as legends, often invented to fulfill Old Testament prophecies. Charming legends, but without historical verisimilitude."

"I feared as much. Well, come on, Jarvis, out with it. Do you believe in God?"

"No, sir. Oh, I should have mentioned it before, sir, but Mrs. Gregstead telephoned."

I paled beneath the t. "Aunt Augusta? She isn't coming here?"

"She did intimate some such intention, sir. I gathered that she proposes to prevail upon you to accompany her to church on Christmas Day. She took the view that it might improve you, although she expressed a doubt that anything could. I rather fancy that is her footstep on the stairs now. If I might make the suggestion, sir . . ."

"Anything, Jarvis, and be quick about it."

"I have unlocked the fire escape door in readiness, sir."

"Jarvis, you were wrong. There is a God."

"Thank you very much, sir. I endeavor to give satisfaction."

Starry, Starry Night

Phil Plait

When I was a kid, I used to have a real problem with Christmas.

It's true. These feelings took root in those deep, dark recesses of childhood where my memory is now dimmed, but I suspect it all started because I was raised Jewish. No doubt some jealousy was involved—I do remember trying to tell my friends how much better Hanukkah was than Christmas because it lasted eight days and not just one—but I suspect it was also just getting sick and tired of constantly hearing about something in which I wasn't participating.

I'm also pretty sure Christmas music had something to do with it. Man, I still hate Christmas music.

So of course I was teased a lot by the other kids. I grew up in a suburb of Washington, D.C., and while there were many Jewish families, we were definitely a minority. Most of my friends were Christian, and in the days leading up to the end of December, Christmas was all they could talk about. I never believed in Santa no matter how much they tried to persuade me of his existence. That made me a bit of an outcast, of course, but I took some consolation in being right.

Over time, things changed, as they tend to do. I was never that big on anything in the Jewish religion, even when I was very young. By middle school I was for all practical purposes an atheist . . . and I suppose that has never changed since, come to think of it. But despite that, my attitude toward the holiday season evolved.

In secondary school, my best friend was Marc. His family was kinda sorta Jewish (the father) and some flavor of Christian (the mother), and they had long since decided to celebrate Christmas every year as a family event. Marc and I were pretty close, so I was over at their house a lot, including at Christmastime. When the holiday approached I would help them get their tree, set it up, string the beads or fake cranberries or whatever the heck they were—I remember one year we tried popcorn, but were less than successful getting it to stay on the fishing line—and then decorate it.

On the night before Christmas, my non-Christian house would be business as usual—dinner, fool around, read, whatever. Even then I was the budding astronomer, so I might take out my telescope for some relaxing, but frigid, sky viewing. But eventually I'd go to bed, unhappy that every freaking TV and radio station (this was long before the Web, kiddies) was either playing the dreaded jingles or simply off the air.

Once I was up in the morning the long wait would begin. I knew Marc and his brother, Dave, would have been up early, opening presents, getting all kinds of awesome gifts. One year they both got Nikon cameras; we were heavily into photography then, with my bathroom at home being a makeshift darkroom complete with noxious chemicals that my mom was always giving me grief over. The Nikon camera Marc got was really nice, much better than my crappy Konica . . . but no, jealousy wasn't an issue then. Of course not.

Finally, after a torturous wait, Marc would call and invite me over, and I honestly had fun sharing in their celebration. His mom would make a Yule log cake (oh boy, do I remember the glare I got from her when I said it looked like a giant Hostess Ho-Ho), and we'd eat tons of chocolate and then go outside in the snow and have fun.

So for a while Christmas was really cool. Of course, in high school I was a band dork, and that meant every December concert I played Christmas music. So the barely restrained murderous impulse was still there, but mollified a bit.

In college, things died down somewhat because all the other students left to go home and be with their families for the holiday. It was great for me because I could stay behind and make good use of all the fallow computers. My software, written to analyze and model astronomical data, ran scads faster since the machines were otherwise idle. I always got a huge amount done during those weeks.

But it was lonely.

With one exception, for a few years Christmas was neither a joy nor a drag. The holiday was just something that happened, a few weeks of sales at the stores, barely tolerable jingles over half-shot speakers at the malls, and half-price chocolate bars the day after the holiday. The one exception that stands out was spent studying for my Ph.D. qualifying exams. I was home with my parents, but I hardly saw them; I was up every night until 2:00 or 3:00 a.m. studying and doing endless exercises in calculus, physics, and astronomy. That particular holiday is now blurry in my memory, difficult to distinguish from fiercely complicated equations, dozens of pages of algebraic computations and notes, and endlessly having to sharpen my pencils.

But this too did pass. As did I, as far as my exams went. But it wouldn't be the last time I would associate Christmas and astronomy.

To me, when I was younger, winter months always meant crisp, clean air, the sharp pinpoints of stars in the sky undimmed by the East Coast's summer haze. In December especially, while my friends were dreaming of gifts and fun, my thoughts would turn to the brilliant colors of the stars in Orion as the constellation stood solidly over my southern horizon. I read everything I could about astronomy, and also practiced what I read: I would haul my 175-pound telescope to the end of the driveway and, shivering in the subfreezing temperatures, patiently aim it at various objects in the sky. Jupiter, Venus, the Orion Nebula . . . these all became my friends as I spotted and studied them.

It was around that time of my life when it dawned on me that people generally misunderstood astronomy. I myself was a victim of this; when I was of a certain age I believed in all manners of nonsense, including UFOs, the Bermuda Triangle, and astral projection. (Well, I didn't actually believe in that last one. Even then I was a budding skeptic and decided to do some experimental testing: I tried to project my mind using a book I found at the library. But, sadly, the girl I had a crush on showed no signs the next day that I had spent an hour trying to communicate with her from a higher plane.)

The more I read about astronomy, the more instances I found of people misapplying it. Horoscopes were hugely popular, of course, as was the idea of aliens having visited humans, teaching us how to draw really long straight lines in the desert and paint confusing imagery on our stone walls.

And, of course, every year in December, the newspapers would have articles about the Christmas star. You know the story: a star appears in the sky to guide the three wise men to the birthplace of Jesus. From the King James Version:

> Now, when Jesus was born in Bethlehem of Judaea in the days of Herod the king, behold, there came wise men from the east to Jerusalem,
>
> Saying, Where is he that is born King of the Jews? for we have seen his star in the east, and are come to worship him.

A lot of folks in America like to interpret the Bible literally, so this passage is clear enough: an actual new star appeared in the sky that guided the wise men to Jesus. Ignoring for a moment that if they lived to the east and followed the star to the east, they'd get further from Bethlehem rather than closer, and that while Matthew makes a big deal of the star, Luke doesn't even mention it—which already makes a literal interpretation of the Bible somewhat dicey—what we have here is an obvious astronomical tie-in with Christmas.

It's a star, after all.

And that means that astronomy is once again intertwined with Christmas. Even as a lad I could see the implications of this story, and certainly every Christmas special on TV has some variation of a brilliant star in the sky as a symbol for Christmas. Really, my getting involved once again with Christmas was unavoidable.

So I thought this legend over. Was the star real? A lot of people thought so. That meant I had to look at the evidence.

It's thought that the wise men were astrologers, so they would've had some familiarity with the sky; back then astronomy and astrology were pretty much the same thing, even if today they are as different as real medicine and homeopathy, or stage magicians and psychics, or . . . well, you get the point.

The point is, these guys would know the sky pretty well. If we take the star at face value, then it must've been something amazing, because these three guys wouldn't have dropped everything to make a long trek over some mundane star. The obvious conclusion is that it must have been very bright.

What astronomical objects are bright, can appear in the east, and disappear after some amount of time?

While there are lots of potential candidates, to an astronomer the answer is obvious: a supernova—a star that explodes at the end of its lifetime—is a perfect fit. So all we need to do for proof of this idea is to look for a 2,000-year-old supernova remnant, the expanding gas from such an explosion.

And lo, some do exist! But it turns out they wouldn't have been in the east, or wouldn't have been bright enough. Certainly none fits the story well enough, and it's doubtful any would've been enough to suddenly inspire a trio of men to get a hankering for a road trip in the desert.

If it wasn't a supernova, then what was it? Another bright astronomical event is a conjunction, when two planets pass near each other in the sky. Jupiter and Venus are both astonishingly bright, and when they pass very close to each other would make a spectacular scene. And they can also both be in the east!

Were there any conjunctions like that around that time?

In fact, there were. Recently, an astronomer, using computer programs to map the positions of the planets in the sky, discovered that in 2 BC Venus and Jupiter passed very close to each other; so close in fact that to the eye they would have appeared as a single star! So have we found the star?

Not so fast. First off, the planets move relative to each other, so even two days earlier or later they would've been seen as two separate objects. The wise men would never have mistaken that for a single star.

And oh, did I mention this apparition occurred in June? The wise men certainly took their time getting to Bethlehem!

Okay, so a planetary alignment doesn't fit our biblical bill either. And you can keep looking for other objects that might represent what the wise men are claimed to have seen, but at some point I think you have to realize that you're grasping at cosmic straws. No real cosmic event matches the description in the Bible well enough to inspire the story.

And yet people keep looking. In December every year, without fail, some newspaper article breathlessly reports some astronomer has found another candidate for the star, what turns out to be yet another weak explanation for a biblical passage of dubious reality.

And every year I read these articles and wonder, why do they try so? What are these people really searching for?

In 1992, as I could just start to spy the Ph.D. lurking murkily at the end of my graduate career, I started dating Marcella. Two years later I had my degree and a job, and the next year Marcella and I were married. After a decade or more of no real religious involvement, I found myself with a Catholic family, one that really celebrated Christmas every year. Food, the tree, midnight mass, reading "The Night Before Christmas," and, yes (sigh), singing the dreaded carols. A year after Marcella and I married, our daughter was born, and that cemented the celebrations: in my family Christmas is absolutely for kids.

Now, it's not like I jumped right into this. Thirty years of secular winters is more than just a habit. At first I was reluctant to participate much. And in some ways this new rekindling reawakened the reasons I didn't like it all those years before.

But then something funny happened: one year I decided I liked the tree.

It was cool. I had a tree in my house. Pine trees smell good. They're pretty. Hanging ornaments and lights, if done properly, are actually rather festive. And I found I liked going out and physically getting the tree. We even once went to a huge farm where trees were grown specifically for the purpose, and I cut one down for us using a bow saw and everything. It was very macho.

Ironically, my wife—raised with this holiday—prefers fake trees. But maybe that's because she always winds up doing the decorating (I'm hopeless at it, and likely to set fire to something) and it takes her all day. However, I won't stand for an ersatz tree. Every year we get a real tree and let it make our house smell piney and arboreal.

And, yes, Christmassy.

Now, after many years of celebrating this holiday, I've come to really enjoy it. I know my in-laws well enough to know what kinds of gifts to get, and my own daughter makes it clear what she wants (somehow the video games we get for her are always the kind Marcella wants to play). I always get the same sort of gift from them: a big Toblerone bar (400 grams), thermal socks (my office is cold even in the summer), and various computer doodads and gizmos.

And every year I'm happy. I mean, honestly happy. Some people say the gifts are not the reason for the holiday, but they're wrong: of course it's about the gifts. They're the centerpiece of the holiday; it's about giving them, and having fun getting them, and then playing with them (or wearing or eating them) afterward. And not to be all TV Christmas special here, but it's about being with family while you're doing all that.

So here I sit. An atheist, a skeptic, a guy raised Jewish who hated Christmas, has found the meaning of the holiday, and he wasn't even searching for it.

And every year, when I read the blogs and the papers and watch the news, I see that same story of the Christmas star resurrected, an undead story that won't stay down. And people keep looking for the evidence.

But they won't find it. They can't. It's a story.

So for me, just being with family, enjoying their company, is enough. And, of course, every winter I still go outside to observe the sky and look at the stars, the real stars. You don't need to search for them—they're there, festooned across the sky for everyone to see.

The Ironed Trouser:

Why 93 Percent of Scientists Are Atheists (Depending on Whom You Ask)

ADAM RUTHERFORD

Atheism and science should make good, comfortable, spooning bedfellows. Even though they are totally separate types of thing, the former being a position, the latter a process, the casual assumption is that they should skip hand in hand through gloriously evolved fields of reason. Those who attack either or both like to conflate the two for a convenient jab-swing combo to pulverize rational thought in favor of religious fervor. Science must be bad because it lies so comfortably with godlessness.

The term "scientific atheism" is tossed around sometimes, but I don't really understand what it means. Atheism exists fully independently of science. As the onus is on the faithful to demonstrate the existence of Yahweh, Allah, Thor, Hanuman, or whomever, atheists need to do nothing at all to be devoted to their stance. "Scientific atheism" is equivalent to saying "ironed trousers." Like science, ironing is a process, which can be applied to all manner of items: dresses, shirts, even underpants, if one were so inclined. It straightens things out, makes them fit together nicely. Fortunately, trousers exist and function perfectly adequately without ironing.

And atheism exists without backup from science. But science does make it look a bit smarter.

In the twentieth century, there were several attempts to quantify the overlap of eggheads who were godheads. In 1916, psychologist James Leuba found that out of 1,000 scientists, 60 percent were agnostic or atheist. Eighty years later, the experiment was repeated, and the results were virtually identical. Within a different sample, only 7 percent of the members of the American National Academy of Sciences indicated a belief in God. More recently, a survey of the fellows at the UK's most august scientific body, the Royal Society, revealed only 3.3 percent who believed in God.

As with so many surveys, it depends on whom you ask and how you phrase the question. Is the Royal Society a representative sample of scientists? Oh Lord, Mary Mother of Jesus, heavens to Betsy, Christ on a bicycle, no. For starters, only 5 percent of Royal Society fellows are women, something like ten times lower than in the general scientific community. A recent survey indicated that Royal Society fellows are 38 percent grumpier than other scientists.* Many fellows are so old it's difficult to ascertain if they are even alive, let alone God-fearing. It is possible that this gives them an inside track on the big answer, but one would have to untimely wrest them from their peace to find out.

But what is clear is that those of a science bent are more likely to also lack religious faith. Why should this be? Because the process by which scientific knowledge is revealed is one that requires logic and rational thought at every stage. Any researcher will tell you that there are plenty of moments that necessitate creative guesswork, or simply having a wild stab in the dark, but in general these moments are massively outnumbered by the grinding out of small incremental steps

* This survey was entirely made up, by me, for the purposes of making a glib point.

toward better theories. Science as a way of acquiring knowledge certainly predisposes one toward ruling out the inconsistencies and irrationality inherent in religion.

Furthermore, science explains how things are. There is a nonsensical variant of the argument from ignorance referred to as the "God of the gaps." Very simply, where there is a hole in knowledge, insert God as the explanatory force. It's nonsensical because historically, it was gaps all the way down. What science does very well is fill them in. To those gappists, I say that just because you don't understand something, doesn't mean I can't.

So there are two robust reasons why scientists are less likely to be religious. But a much more interesting question is why any scientists are religious. Opponents sometimes screech that scientists have to have faith in science itself. This is true in a sense, but at least the robustness of the scientific method is such that a belief that the system works is based on countless data points that show it to be reliable: where once there was ignorance, science has inserted knowledge. Having faith requires an absence or ignorance of scientific evidence, a belief that is not supported by a logical progression. That's why it's called faith.

One might be tempted to suggest that scientists who believe are not very good scientists. Empirically this is simply not true, and I'm not talking about the preachers of that creationist fig leaf they call intelligent design. No, there are plenty of good scientists who are religious, who have faith, who see the laws of nature, evolution, gravity, the whole damned universe as a manifestation of a non-interventionist divine force that now acts like an absentee landlord: he sets up the rules of the cosmos and then clears off forever. These people are technically deists.

I don't really see the point of this stance, but I accept that the cultural trappings of religion can be hard to shake. It may be one of my own bountiful shortcomings, but I have not stumbled across a convincing argument for this apparent internal conflict that doesn't rely on a form of compartmentalization of one's rational and irrational minds.

And that's fine. Everyone, even the most hard-line rationalist, behaves in absurdly irrational ways. It's the nature of humankind. I couldn't believe in God any less: it makes no sense to me, and more importantly, my trust in science's extraordinary explanatory abilities renders the need for divine answers superfluous. All things are potentially explainable without recourse to the supernatural. But that doesn't mean I exist in a purely rational way. I've spent the past twenty-eight years supporting a football team who in that time have won a grand total of two trophies, both before I was seven. All because of the random cosmic happenstance of having emerged into the world in a hospital lift in the small market town of Ipswich. And even so, I will be a "tractor boy" till my cardiac myocytes twitch their last. Is that rational? No. It's not even very much fun much of the time, goddammit.

While some consider it to be a weakness, the true strength of science is that it is always and willingly subject to being wrong. A scientific truth that is right today may yet prove to be incorrect, or need to be modified in incremental steps toward a better, truer truth. If the supernatural turned out to be real, with God and angels and demons and unicorns and behemoths and whatever else, then it would instantly stop being super- and start being just natural. At that point, scientists would want to know what the hell was really going on.

I like to fantasize that God does exist, and what He and I might talk about. In the extremely unlikely event that He did appear before me, it would indeed be a revelation. Who knows? I might even indulge in a bit of glossolalia. But once I'd reassembled my lower jaw, stopped gibbering, composed myself, and apologized to my devout Catholic gran for giving her such a hard time all these years, the realization would be that although much of what we assume to be true is not, the revelation would simply open up a new, mouth-breathingly exciting branch of science. If He did make everything, quarks and all, then surely he'd be pretty excited to let us mortals make some new discoveries:

Me: Sorry about all that ardent non-believing I've been doing. By giving me choice, you didn't really give me much choice.

God: Don't sweat it. Any questions? I'm in a bit of a rush, I've got an urgent dice game to play with Einstein.

Me: Right. Did Maradona handle the ball in the 1986 World Cup finals against England?

God: "Hand of God," my divine arse. Nothing to do with me, mate.

Me: I forgive you. Listen, loads of questions to ask you, like "What have you been up to for the last 13 billion years?" and "What the hell is the point of Belgium?" But I'm just going to stick to the facts: What are you made of?

God: Well . . . [answers in full]*

Me: Riiiight. Wow. That explains why in 10,000 years of history we haven't been able to categorically verify one single instance of Your existence.

God: Yeah, sorry about that.

Me: We're gonna need some new technology and a seriously colossal grant to start researching this.

God: Anything else?

Me: One last thing. Would you mind just clearing up the "Thou shalt not kill" commandment? There seems to be a bit of confusion about it here on earth.

God: [slightly embarrassed mumbling, exit stage left]

---

* Precise details of this section of this conversation were unavailable at the time of going to press.

One can but dream. What scientists are very good at is asking questions. The scientific method provides a framework that allows us to ask those questions, rather than accept assertions. Take the example of the great and never-ending shouting match between those who understand evolution and those who are unencumbered by the gifts of fact or reason: creationists. A literal interpretation of the biblical account of creation fails at every possible rational or scientific question one might put to it. It is an assertion of truth based on nothing other than a fiction.

A thousand years ago, it wouldn't have been all that easy to demonstrate how creationism is wrong. It existed largely in a knowledge vacuum, devoid of any evidence to the contrary, or any understandable evidence at all. The age of the earth was unknown, the fact of evolution was unobserved, and the idea of a high-throughput automated fluorescent DNA sequencing machine was a matter for the dunking stool. For almost every question one could ask, the answer would be "We don't know." For many years biblical creation was the only explanation. How were they to know that snakes have almost none of the physical attributes required to talk?

But by the first half of the nineteenth century, well before Charles Darwin graced us with evolution by natural selection, plenty of evidence had accrued that indicated that creationism could not be right. A whole steaming heap of wrong. And then in 1859 Darwin published the *Origin of Species*. In it he outlined one big idea that not only fitted the observed evidence about the age of the earth and the process of evolution but also made predictions about what we would find next, many of which turned out to be very right. It's not so much that creationism is wrong (which it most certainly is) but that that evolution by natural selection is so much righter. So right, in fact, that it is now the only sensible way of understanding the origin of species on Earth. With varying degrees of wrongness, other ideas and theories have come and—via the bypass of experiment and the slip road of failure—gone. It now seems unlikely that any theory will come along that could replace natural selection wholesale. But should that happen, scientists would be committed to investigating it fully. Currently, and for the foreseeable future, evolution by natural selection is categorically, emphatically, and by far the best explanation for understanding the breathtaking diversity of life on earth.

Evolution, as a scientific fact, is nothing much to do with being an atheist. It has a lot to do with ruling out medieval religious dogmas as childish hangovers from an ignorant past. But the process by which evolution was realized, tested, and modified has a lot to do with the revelation of knowing that there is probably no God.

And that is science's greatest strength: as a way of knowing. It's an unending pathway toward knowledge and enlightenment about how stuff works. It's a thought process based on observation, experimentation, rational thinking, and logic. There's no recourse to jumping to conclusions or leaps of faith. There are dogmas in science, but they are always subject to change. When it's wrong, it's wrong, and we need to modify our preconceptions and develop a new and better way of tackling the problem. That's why science is the best way of knowing how things truly are. And as such, it's a way of thinking that should have the effect of eroding faith. So whatever the real number of egghead godheads is, the fact that there are any at all reveals not a weakness of science nor a strength of religion but the fallibility of people.

The Large Hadron Collider:
A Scientific Creation Story

BRIAN COX

The Large Hadron Collider (LHC) at CERN in Geneva is the biggest and most complicated scientific experiment ever attempted. More than 10,000 scientists and engineers from eighty-five countries have built a machine that can re-create the conditions present in the universe less than a billionth of a second after the Big Bang. The reason that the world has come together at CERN in the pursuit of pure knowledge is simple: we want to understand how the world came to be the way it is. This quest has led to a remarkable description of the violence and beauty of the origin of the world, and ultimately the emergence of life and civilization in our universe.

Around 13.7 billion years ago, something interesting happened, and our universe began. One ten-million-billion-billion-billion-billionths of a second later, gravity began to separate from the other forces of nature and has remained a weak enigma ever since. After a billion-billion-billion-billionths of a second the universe underwent an exponential expansion, growing from less than the size of an electron to the size of a melon in one-hundred-thousand-billion-billion-billionths of a second. The universe then steadied its growth, and the energy that drove the expansion was transformed into subatomic particles, the building blocks of everything in the universe. Around a million-millionths of a second after the interesting event, something known as the Higgs field began to behave in an unusual way. This caused most of the subatomic particles to acquire mass, and there was substance in the universe for the first time. From this point onward, we are reasonably sure that our story is correct because over the past century the LHC's smaller cousins have explored these violent conditions in exquisite detail. We are therefore the first culture in history to engage in a program to test our creation story experimentally. The primary job of the LHC is to explore the story during the time when the Higgs field became influential.

The LHC is a 27-kilometer-long circular machine that accelerates subatomic particles called protons to as close to the speed of light as is possible with our current technology. Approximately half of your body is made up of protons; the other half is made of neutrons. The machine straddles the border between Switzerland and France, which the protons cross 22,000 times every second inside two parallel drainpipe-size tubes. More than 1,600 powerful electromagnets, operating at –271 degrees Celsius, keep the protons spiraling neatly around the machine in precisely controlled orbits. The tubes cross at four points around the ring, allowing up to 600 million protons to smash into each other every second at each point. Surrounding these mini-explosions are four detectors: digital cameras sitting inside cathedral-size caverns 100 meters below the vineyards and farms. It is their job to photograph the stage in our creation story that we want to explore.

According to theory, the Higgs field acts like cosmic treacle. The subatomic particles that make up our bodies and everything we can touch in our world acquire their masses by interacting with this all-pervasive stuff. Imagine attaching a string to a ping-pong ball and pulling it through a jar of thick treacle. If you didn't know better, you might conclude that the ping-pong ball was very massive because it feels difficult to move. This is roughly how the Higgs field works in our best theory of the subatomic world, known as the standard model of particle physics. It may sound far-fetched, but the Higgs model has survived for more than forty years without actually being shown to be correct because it has very elegant mathematical properties that physicists find convincing.

With the LHC, however, D-day has arrived for the Higgs model. If it is correct, then particles associated with the Higgs field, known as Higgs particles, must show themselves in the LHC's underground detectors. We can be so sure because, to do the job necessary in our creation story, the Higgs particles must be light enough for the LHC to create them in its high-energy proton collisions. If the Higgs particles don't show up, then nature must have chosen some other mechanism to generate mass in the universe, and we will observe that instead. It's as if the LHC allows us to journey back in time to the point in our story where mass appears in the universe for the first time and take pictures of this most important of historical events. Because we can repeat the collisions billions of times, we can carry out very high-precision measurements that will allow us to investigate our creation story scientifically.

This time in the universe's evolution is known as the electroweak era, because two of the four forces of nature, the familiar electromagnetic force and the less familiar weak nuclear force, reveal themselves as different facets of a single unified force at these temperatures. The weak nuclear force is shielded from our everyday experience deep within the atomic nucleus, but it is vital in allowing the sun to shine because it allows protons to change into neutrons, and therefore hydrogen to fuse into helium with the release of sunlight. The LHC will probe this unification, which intimately involves the Higgs mechanism, with unprecedented precision and verify or refute our current theoretical models.

There are also hints that there may be surprises in store. Some particle physicists believe that the standard model Higgs theory is flawed because it requires a very delicate fine-tuning of parameters to make it work. Fine-tuning is considered ugly in physics; if the universe only works if the strengths of the forces or the masses of particles take on very precise values, then physicists naturally want to know why this should be so. Coincidences do happen, but it is wise to look for more elegant explanations. There is a popular alternative to the standard model that goes by the name of the minimally supersymmetric standard model, or MSSM. This theory requires a doubling of the number of fundamental particles in the universe, plus no fewer than five different Higgs particles.

This sounds like additional complexity rather than an elegant simplification, but the MSSM achieves more than solving some of the fine-tuning problems: it also provides a possible answer to a decades-old problem in astronomy. It has been known for some time that there is much more matter in the universe than can be accounted for by simply counting up the number of stars and galaxies that we can see. In fact, it appears that five times as much matter is required to explain the orbits of stars around galaxies and the motions of large clusters of galaxies through the universe. Models of this missing stuff, known as dark matter, work best if the missing matter takes the form of an as yet undiscovered heavy subatomic particle. Within the MSSM, such a particle does exist, and if the model is correct, then this particle and a whole new zoo of its sisters should show up at the LHC. Such a discovery would represent a giant leap in our understanding of the subatomic world and the evolution of the universe as a whole.

From this point onward, we move into the realm of the scientifically well tested, and our story can be told with more certainty thanks to the generations of particle accelerators that went before the LHC and decades of study in nuclear physics, cosmology, and astronomy. After a millionth of a second, the four forces of nature had taken on the separate identities we see today, allowing a sea of particles called quarks and leptons to interact in a dense subatomic soup. After a second or so, the universe was cool enough for the quarks to stick together into protons and neutrons, and particles called neutrinos were freed from the soup to roam through the universe forever. There are several hundred of these primordial relic neutrinos in every square centimeter of space today, including the space inside your body. Three minutes passed, and protons and neutrons began to form simple chemical elements. In less than half an hour, the universe's supply of hydrogen and helium was fixed in the ratio we see today: 75 percent hydrogen to 25 percent helium.

For another 380,000 years the universe was too hot and dense for light to travel through, but as it continued to expand it reached transparency and the photons from the violence of the early years were set free. We can detect these photons today as an ever-present visual hiss known as the cosmic microwave background—a fossilized picture of the early life of our universe.

The universe remained locked in a cosmic dark age until the weak but ever-present force of gravity began to cause clouds of primordial hydrogen and helium to collapse, forming the first generation of stars. These stars fused hydrogen into helium and, in a complex dance between the strong and weak nuclear forces and electromagnetism that relies on an incredibly delicate balance between their relative strengths, helium stuck together into longer chains; three helium nuclei glued together make carbon, the element of life. As the first generation of stars ran out of fuel, they exploded, scattering the newly formed carbon, oxygen, and other light elements into the universe. During the explosion itself, gold, silver, and the heavy elements were made.

The interstellar clouds, now enriched with the building blocks of life, collapsed again under the inexorable influence of gravity to form stars with dense, rocky planets orbiting around them. On at least one of these planets, the newly minted elements got together to form complex structures capable of self-reproduction, and the universe sprang into life. On this precious world, single-celled organisms began to cooperate in colonies that, given billions of years of relative stability on the surface of the planet, worked out how to journey through space and, in the year they arbitrarily called 1969, left the imprint of a structure they called a foot on another world.

There are those who argue that science removes the majesty from the universe by demystifying it. My reply is that the scientific creation story has even more going for it than the virtue that it is most likely correct, at least in its broad sweep. It teaches us that we are part of nature, built of the same stuff as stars, planets, asteroids, and comets. Our protons and neutrons have been around since the earliest times, glued together into heavy elements in the nuclear furnaces of long-dead ancient suns, blasted out into the universe and resculpted from diffuse interstellar dust clouds by the gentle hand of gravity. We are colonies of particles that have learned to think; every human is a grand natural structure, an emergent form permitted to exist by the laws of nature and realized by a stream of coincidence and causality. When the pattern of atoms known as you ceases to be, the building blocks will return to the voids of space, and in a billion years or more they may take their place in another structure so beautiful that a future mind may perceive it to be the work of a god.

The scientific creation story has majesty, power, and beauty, and is infused with a powerful message capable of lifting our spirits in a way that its multitudinous supernatural counterparts are incapable of matching. It teaches us that we are the products of 13.7 billion years of cosmic evolution and the mechanism by which meaning entered the universe, if only for a fleeting moment in time. Because the universe means something to me, and the fact that we are all agglomerations of quarks and electrons in a complex and fragile pattern that can perceive the beauty of the universe with visceral wonder is, I think, a thought worth raising a glass to this Christmas.

How to Understand Christmas:

A Scientific Overview

NICK DOODY

"Christmas"—it's one of those words we bandy about confidently every day. "Happy Christmas!" we shout at children in the street. We send cards inviting friends to "come Christmas with me." Anyone lucky enough to find himself in Cardiff on February 29 has surely joined in with the chorus of "A-Christmassing Down the Mound." But could any of us truly say what the word means? Tests have suggested that we could not.*

The scientific history of Christmasology is a rich and fascinating one. My purpose here is to give an overview accessible to the layman, but not so simplified as to misrepresent the true story. As ever, it is impossible to please everyone, so I shall take the opportunity to apologize to readers hoping for an in-depth explanation of Barsky's chimney hypothesis; there simply wasn't room. Conversely, if newcomers to the field find themselves unfamiliar with terms like

---

* Pennheisner, 1954. See also, "Have We Forgotten to Remember Not to Forget that We Never Knew the True Meaning of Christmas?" *Daily Mail*, December 12, 1999–present.

Sado-Melchiorism or *tree*, I hope that they will forgive me, and make use of the bibliography to dig deeper into this fascinating seam of study.

The Christmas story with which most of us are familiar begins in 1858, in Berlin, where a young philologist and psychologist, Bernhard Gernhard, began to question whether we all meant the same thing when we said "Christmas" (the story that his brainwave came in a dream as he slept on a stuffed reindeer is an entertaining one for children, but probably apocryphal). The thrust of Gernhard's initial train of enquiry was as follows:

We all accept that certain things are Christmassier than others.

We largely agree on what those things are.

Therefore, he reasoned:

We should be able to quantify "Christmassiness," because Christmassy things should have it in common.

To modern Christmasologists, Gernhard's initial experiments might seem amusingly naive, but they are also recognized as germane to the discipline as a whole, and without him the twenty-first-century world would look very different. The North Pole, for example, would still be a barren, frozen wasteland without the magnificent Yule Hypercollider we take for granted today.

As so often, we learn more from Gernhard's failures than his successes. Almost everything we know about the Christmassiness of beards, for example, can be traced to his experiments in the spring of 1860. Before then, it was generally considered that beards themselves had a Christmassy quality when viewed from the lap of the beard's owner. Gernhard had no initial reason to dissent from this view, but his attempts to measure *Bartweihnachtheit* threw up some interesting and unexpected problems.

Gerhard's first experiments involved placing a child at varying distances from the lap of a bearded man and noting how Christmassy the child reported feeling.* Naturally, he expected Christmassiness to increase with proximity; the point of the experiment was to find by how much. What no one could have expected, though, was that some of the children, even placed directly on to the lap of a bearded man, didn't feel Christmassy at all.

Gernhard was dumbfounded. He recalibrated his experiments, double-checked the quality of his children,† and tried again, first approaching the bearded men's laps at a painfully gradual pace, next practically firing the children at the laps. Still some of the children reported no Christmas. Gernhard's quest seemed to be at a dead end.

It was three weeks before it dawned on Gernhard that the fault might not be with the children. He was putting together a new team of bearded men‡ when it suddenly occurred to him: what if not all beards are Christmassy? Some quick thought experiments convinced him that he was on to something: it had been assumed by the Greeks that, say, a sea urchin with a beard isn't Christmassy because the un-Christmassy nature of the sea urchin cancels out the Christmassiness of the beard. But no one had ever considered that some beards

---

* No really accurate way of measuring Christmassiness was developed until the 1970s, and Gernhard's results are riddled with false positives from children approaching a birthday, wearing new sweaters, etc. It was also not yet understood that orphans have a naturally high level of Yule, a factor that today's researchers adjust for statistically.

† A true product of his time, Gernhard's first thought was that some of the children might be ghosts, but close questioning of the parents ruled this out.

‡ The original bearded men were mainly Orthodox Jews, and it was felt that a fresh batch might make the experiment more pleasant to conduct. Gernhard's notebooks record that "the Jews are becoming irritable," and the atmosphere worsened by week two, when the purpose of the experiments was explained to them.

might not be Christmassy at all! No one had ever before attempted to imagine something potentially Christmassy with a non-Christmassy beard. Before 1860, such a thing was literally unimaginable. Close your eyes for a moment and picture yourself being given a plate of sprouts by a female horse with a blue beard,* and you might come close to understanding Gernhard's excitement.

Over the next eight years, Gernhard's experiments refined and expanded his theories, and in 1868 he published his collected observations in *On the Christmassiness of Beards*—a true milestone in science, famously referred to by Mark Twain as "a Christmas *Origin of Species*."† It won him a place in history and the inaugural Templeton Peck Award for Christmas Science.

What Gernhard had achieved was a complete shift in the scientific consensus, akin to the Chomskyan revolution in linguistics or the Wang theory of bendy numbers. Christmassiness was no longer seen as an essential quality pertaining to an object, but understood as the subtle function of a series of variables. *On the Christmassiness of Beards* not only taught us that white beards are more Christmassy than black ones, but also that this Christmassiness can be decreased by the addition of sunglasses or Chineseness, or increased by jolliness or sitting in a cave.‡ Not for nothing is the nineteenth century known as the Christmas century. From being a sideline interest, the pursuit of a few moneyed eccentrics, Christmasology now took its rightful place as a "proper" scientific discipline, sitting proudly alongside biology, space physics, and clapometry.

* Not a white beard, as that will make the image Christmassy.

† Not to be confused with the Tim Allen comedy movie of the same name.

‡ As an aside to demonstrate a real-world application of Christmasology, the cave variable dogged Osama bin Laden throughout his later years. No matter how threatening his rhetoric, once his beard had started to gray the image of him in a cave would make viewers of his videos feel Christmassy—almost the exact opposite of his intention.

One might be forgiven for thinking that Christmasology took a backseat to war in the first half of the twentieth century, but on at least one occasion in 1914, the opposite is true, when soldiers in the trenches ceased fire so that keen amateurs on both sides could examine a Christmas that had landed in no-man's-land. Still, war and its aftermath took its toll, and when the Second World War began, making hideous sense of the name of the First World War, many Christmasologists were put to work cracking the Enigma code or helping build the atomic bomb.

The second half of the twentieth century saw Christmasology explode as a discipline, diversifying into subfields like Geochristmasology and Medicochristmasology,* and even sub-subfields like Dermachristmasology.† Of these, perhaps the most interesting from the historical point of view is Christmasozoology.

Huge strides have been made in the past fifty years in understanding the intricate relationship between Christmas and nature. Barsky's chimney hypothesis (BCH), too complex and difficult to go into here, or indeed anywhere, proved the missing piece of the puzzle when it came to discovering the evolutionary advantages gained by reindeer that spend part of the year on roofs. For his integration of BCH into reindeer game theory, Christmasomathematician George Maynard Carol was rightly awarded 1968's Templeton Peck Award.

Even more exciting were the discoveries of the 1970s, still known as the Christmas decade. By now it was accepted that Christmas occurred in nature, and that certain animals (reindeer, camels, oxen—especially with assen) are naturally more Christmassy than others

---

* Not to be confused with Christmasomedicology, which is a branch of medicine.

† Not to be confused with Christmasodermatology, which is a subfield of Christmasomedicology.

(sharks, minotaurs, robots).* But in 1974, Carol George and her husband, George, showed that it was possible to intervene and artificially elevate the Christmassiness of an animal.

The Maynard effect, as it is now known (Carol called it after her middle name), is observed if one compares a robin with some mice. Robins, as is well known, have an unusually high level of natural Yule—far higher than a mouse. In fact, even a hundred mice are not as Christmassy as a single robin. What the Georges demonstrated, though, is that you can approach robin-like Yule levels with only about fourteen mice by dressing them in tiny waistcoats.

The Georges' work was greeted with interest but did not, at first, overwhelm. For one thing, they were still measuring Yule levels with the modernized Gernhard method—essentially, asking children how Christmassy the test subject made them feel, then adjusting statistically for orphans. This was no indictment of their experiments; the MG method was the best available at the time, but it made the scientific community reluctant to greet new Christmas research with anything but a cautious welcome.

Then, in 1978, something incredible happened: Donald Maygeorge invented the Christmasometer.

Maygeorge, a naval engineer by background with only a layman's knowledge of Christmasology, was tinkering about with some naval tinsel in his shed one dull 1978 afternoon when an idea suddenly hit him. And it was an idea that, had he been named Newton, would have made him the second most famous Newton† in science.‡

What Maygeorge suddenly realized, and what no one else had thought of, was breathtakingly simple: by taking a Van Rijd detec-

* Except Christmas robots, of course.

† Probably after Isaac Newton.

‡ *I Smell Christmas: The Donald Maygeorge Story*, by Colin Harper.

tor (the main component of a clapometer), isolating its feedback loop, and replacing the Fenchurch plate with a Bethlehem coil,* you suddenly have a Van Rijd detector that works with tinsel instead of compère's rouge, and uses the tinsel to boost its own output wave as a square of the tinsel's Yule levels. A child could have thought of it! Suddenly we lived in a world that had a way of measuring Christmas. Aristotle had been right after all.

From that spring in 1978—still known as the year of Christmas—the whole realm of Christmasology had changed. Not only was there now an accurate way of measuring Christmas, but another implication soon dawned on the Christmas world. The Sezniak hypothesis had to all intents and purposes been proved: there must be a Christmas particle.

More than thirty years on, we know more about Christmas than our ancestors ever dreamed possible. We have identified the Christmassiest color in the spectrum;† we know which is the most Christmassy planet in the solar system (it's Earth);‡ our computers have calculated the human name most redolent of Christmas;§ and anyone can tell a genuine caroler from a singing mugger or "wassailant" at the flick of a switch. But do we really understand what we mean when we say "Christmas"? As I indicated before, it appears not.¶ But we shall soon.

---

* Even a common Walsh repressor will do.

† Wenceslas green, which lies halfway between red and white.

‡ Although Neptune is a very close second, for reasons that puzzle Christmastronomers.

§ It is Hapsail Turkey Scarfnice, a computer-designed name expected to be as popular as Andrew by the year 2023.

¶ Pennheisner, 1954. See also "Have We Forgotten to Remember Not to Forget That We Never Knew the True Meaning of Christmas?" *Daily Mail*, December 12, 1999–present. See also the present article.

The last step on this remarkable journey must be the Christmas particle. The Sezniak yulion is the final part of the puzzle that began with Gernhard's naive beard work. Once the particle is identified, the Christmas mystery will be solved. The quantum event triggered by putting a waistcoat on a mouse, a scarf round an obelisk of snow, or a white beard on a fat, laughing man will at long last be understood. Man will be able to look the robin in the eye and say, truthfully, "I know you."

The Yule Hypercollider, to be activated at the North Pole in just a few short weeks (at the time of writing), is the key to this final step. Those who understand Christmas rather better than the hysterical doom-mongers have, thankfully, overruled early objections that it might destroy the world. Some of the technical difficulties, such as the disembodied, screaming voices, the mysterious faces in the sky, or the strange behavior of animals,* have delayed but mercifully not stopped the hypercollider's progress, and I hope you will join with me and with all mankind on December 25 as we look north with bated breath and wonder in awe as the Christmas story finally comes to an end.

---

* To take two examples, the stag that somehow managed to get into the office of the chief engineer, intone the word "no," then burst into flame, and the penguins that took flight and appear to have left Earth's atmosphere can both probably be explained by unusual electromagnetic fields around the collider.

HOW TO

That it will never come again is what makes life so sweet.

—Emily Dickinson

How to Have the Perfect Jewish Christmas

MATT KIRSHEN

I am more than aware when writing this that, for a Jew, Christmas shouldn't be a time of celebration. In many ways, it was when it started to go wrong—the anniversary of our usurping by the shiny new upstart. A Jew celebrating Christmas is a bit like the family of Samuel Morse toasting the birth of Nokia.

In many ways, Easter should be our time of celebration, and for many years I've petitioned to have its name changed to "We Got Him."

But there is much more to Christmas than remembering the birth of Jesus. There is of course food. Food is so very much the bedrock of Jewish culture, and any occasion built around the dinner table, family, and arguing is as Jewish as . . . Jewish pie?

There is, of course, an obvious problem, one that dogged me all through my childhood. A lot of the best food is non-kosher, so we could not have it in our household. All finest Christmas fare: pork-based stuffing, chipolata sausages, and even the simple act of finishing a meat-heavy dinner with a cream-based dessert (more on that later) are all verboten under the strict laws of Jewish dining. My mother, a good Jewish mother, would never allow such food on our tables, on our plates, in our house. Which is why we went to my aunt's.

There are two ways you can observe a religion. You can either assume that everything in it, as the commandment from an omnipotent, omniscient, omnipresent being, must be followed to the letter, or you can—you know—pick and choose bits that are convenient.

This was, and indeed still is, my family's approach, but no longer mine, I should point out. I've happily shed my faith, and quite frankly, bacon's nice. It's delicious, and I'm not going to deny myself the tastiest meat in the world, at home or away, on the orders of a God I don't believe in. It's like driving below the speed limit on an empty road when you know there are no cameras.

It's odd how hard this is to explain to friends, even ones who know me quite well. It's remarkable the number of times I've been out to eat and someone's told me that I can't have the bacon. Why? "Because you're a Jew. Jews can't eat bacon."

It's as if we physically can't eat it: pork disagrees with us. Moses came down from the mountain with basic dietary advice: "Children of Israel, it's about the bacon. You can . . . but I wouldn't."

My family home is kosher. No pork, no bacon would ever be cooked in our kitchen. We would leave the house and eat it, but never in the house. The house is kosher. As long as the house is going to heaven, that's all that counts. It's important that my parents can stand on their doorstep, gesture backward, and say, "This is a Jewish house. This house has never broken a Jewish law. No non-kosher food has touched this house."

That's not to say no kosher food has been in the house. It's just never touched it. This is where it gets a tad stranger. We would happily order in Chinese food or pizzas—say, pork chow mein or a Hawaiian pizza—but we would eat it off paper plates and protect the table with newspaper. To trick God.

That's not even the only rule as to what makes something kosher.

There are loads of rules, and anyone wanting a properly Jewish Christmas would do well to learn them all, or find a sufficiently hospitable aunt. Mine is available for a small fee. Keen scholars of Judaism (or readers of the beginning of this piece) will know that, as well as the whole pig issue, dairy products and meat products must be kept separate. So, for example, if you're eating a meat-based bolognaise sauce on pasta, you couldn't put cheese on it to make it, well, nice. Cheeseburgers are out too, because they are nice. Essentially the Jewish God hates the delicious.

This separation of foodstuffs stretches to cutlery and crockery too. I remember, as a child, non-Jewish friends getting that wrong: using a "meat fork" for a dairy dish. This fork is now bad; it's contaminated. I'm sure there's a proper name for it, and I could probably look it up, but to save time, let's go with Jew-icky.

If you want to save that fork (and who wouldn't?) your only choice is to bury that fork in the ground for three days, which somehow magics it better. Three days later it's as good as when it was first plucked from the earth all those years ago. As a child I genuinely witnessed my mother, on hands and knees in our back garden, burying a fork like it was the family hamster, which left me thinking, *What the hell are you doing? Surely you either believe in God or you don't. And if you don't believe in a Jewish God, why are you digging for cutlery, and if you do believe in a Jewish God, what were you doing with that BLT five minutes ago?*

So there's your guide to all things kosher. Well, not all things kosher. Not even close. You also can't eat shellfish, dogs, or birds of prey. Anyone longing for a traditional Christmas kestrel would be wise to steer clear of a Jewish household, where the kestrel, rather than being killed and eaten, is kept alive at the head of the table. (I'd like to think this is obviously a joke, but a small part of me really hopes it is taken at face value. Please do your best to spread this as a genuine fact. There's little that would bring me more joy—and Christmas is, after all, a time for joy—than for it to be told back to me in earnest, by someone who claims to have seen it with their own eyes.)

From that moment on, most things can be kept the same. The cracker jokes may have to be changed, as there's too rich a tradition of Jewish humor to be sullied with weak puns on the phrase "mince pies." If you really want to lighten the mood, try replacing them with your favorite Henny Youngman one-liners, and see the delight on your wife's face as you read out what a now-dead twentieth-century gagsmith really thinks of her.

It may seem odd for a Jewish-born atheist to tell you how to enjoy your Christian festival, but why not? It now seems almost trite to point out most of the trappings of Christmas have little to do with the birth of Christ. The tree is pagan, the star is pagan, and the carols are simply a way for schools to keep track of which voices have broken (it's now illegal to check any other way).

Christmas is a festival built on the very nature of picking and choosing the parts that are convenient. So strap on your tinsel, adjust your most Jewish paper crowns, and enjoy your child's Nativity play as you would any other piece of classic theater. Because whatever your faith, Jewish, Christian, other or reasoned, Christmas is about the same thing. A day off. And delicious kestrel.

How to Have a Peaceful Pagan Christmas

CLAIRE RAYNER

It was a blustery, gray December afternoon in the 1970s, and the woman standing beside me at the school gates said, in the way people do when they think small talk is needed, "What are you doing for Christmas? We always start the night before with midnight mass—tiring, but so special, don't you think? Particularly for the children."

Here we go again, I thought, and answered as simply as I could. The children were due out in a couple of minutes, so: "I'm a humanist," I said. "Not religious."

She stared at me blankly for a moment and then said, "Oh, like the Jews, I suppose. Don't believe in the Baby Jesus and his birthday."

"A great many Jewish people are very religious, just like some Christians," I replied, "though some are humanists like me."

"Really." This time there was a clink of ice in her voice. "So what do you people do at this time of year, then?"

It was the "you people" that got to me. "Oh, the same as you do, I imagine," I said sunnily. "Spend too much money, spoil the kids something rotten, eat and drink too much, and fight with each others' relations. We believe in fun and revels and just being together, you see, even if we do spar a bit, just as northern people always have done in the depths of winter—and they've been doing it for much longer than a mere couple of millennia."

And then, at last, the children came cascading out of school and rescued me.

I thought, as I walked home with my whooping offspring playing their "I'm a cowboy" game all around me, about the midwinter revels, and also about man/womankind's great spring festivals, when the sun is thanked for coming back at last and bringing fertility to plants and animals. These festivals encourage humanity to join in a bit of being fruitful and multiplying (if they want to—there is no punishment for those who don't) and recognize the re-creative aspect of sexual activity as well as the creative. (For millennia, people have used contraception in one form or another, just to have fun without progeny. The ancient Egyptians had some interesting ideas about that.)

In many parts of Europe, girls danced round maypoles, which were undoubtedly phallic symbols seen as a little encouragement for those in need of something to get them going. I remember being taught at my infant school in the thirties how to plait ribbons round a maypole as I sang a special Maytime song about bees and flowers.

Meanwhile, poets and authors down the centuries wrote umpteen verses and yearning prose about "maying" and how, in May, it was pretty well a duty to be loving and sexy. Robert Herrick begged his Corinna to come maying with him, and remember Shakespeare's songs?

> In springtime, in spring time
> The only pretty ring time . . .
> Sweet lovers love the spring.

and

Between the rows . . . the lovers lie . . . in springtime.

In his day, by the way, the winter festival was given much greater weight than it is now. It lasted for a full twelve days; hence his jolly play *Twelfth Night*.

To the Romans, the spring festival was called Oestrus, hence Easter in English and Easter eggs, but they made no mention of bunnies. Maybe it is the fecundity of rabbits that makes them a good fertility symbol—and the Christians, cleverly picking up on a very ancient human view of the springtime reappearance (i.e., resurrection) of dead plants and winter-vanished greenery, added on their version of a resurrection myth, telling the story of the magical return of a dead messiah.

Did the lady at the school gates know anything of Norse mythology? Had she heard the glorious tales of a bunch of feisty gods and goddesses and their Valhalla, where dead warriors fought all day and then rose again for another day's bout, and a really cracking festival designed to tempt the sun back when it was its lowest in the sky, because it disappears almost completely as you travel farther north?

Their festival had a logic to it—it was worthwhile for them to make the great effort to follow all the old rituals (it's hard work dragging huge trees through the forest to get them to a bonfire) if as a result they could make the sun do what they wanted it to do. The accompanying tra-la-la, the feasting, the drinking, the swapping of gifts, and above all the sexy games were the reward for their efforts and also a way of showing the sun what great people they were and worth coming back to and how they would repeat the fun when he did come back. As he did, of course, every spring. As primitive forest dwellers, dependent on plants' and animals' fertility and growth for all their needs, their actions seem reasonable enough to me. To worship the sun as the giver of life isn't at all off the map. That is precisely what it does.

It was why they set to work to build and light great bonfires sending hot bright flames leaping up into the dark sky to show the sun in its hiding place what was required of it. It was why they burned the biggest Yule logs they could find (phallic symbol, anyone?), consumed lashes of booze and preserved reindeer meat and salmon and so forth, and then cuddled up into the warm pelt of the man or woman they fancied. (Or, of course, the same-sex person they had fallen for.)

Would the woman at the school gates have been interested in the festival's Oak King, who ran around the forest with great energy, and the Holly King, who liked nothing better than to dress up in red and put a sprig of holly in his tangled hair and then, on the shortest day of the year (December 21), drive eight reindeer and throw a few gifts of food around? A bit like Kris Kringle, who picked up the idea from the Scandinavians. And in America, of course, Coca-Cola invented for their advertisements a modern Santa Claus by putting a jolly bearded chap into a red outfit (originally green, by the way, just like the Holly King) and sending him skyward in a sleigh with—wait for it—eight reindeer.

And what about the symbolism of the Old Religion, as it was called? It is represented today, I understand, by Wicca—adherents of which are described by the uninformed as witches—and modern Druidism, which had (still has) a great admiration for great standing stones, midsummer morning, and mistletoe. The last is because mistletoe is a plant that doesn't trouble to push itself into hard cold ground to thrive but simply perches on the forks and branches of other trees' trunks to tuck into their sap and grow its greeny-gray leaves and those beautiful translucent white berries.

I have wondered whether the Old Religionists long ago decided that the somewhat spermy juices that emerge from a mistletoe berry if you crush it were, yet again, a male sex symbol and gave men permission to kiss under it women who might be inclined to oblige, though they might of course also be those who would much prefer they didn't. But, well, it's Yuletide, so why not have a try?

That is how Wiccans regard the winter revels, too. They are not in the least uptight about sex, being well aware of its value and approving of the joy it brings. Take note, fellow humanists: it's hard to put a fun religion down.

So a pagan version of the "Christian" Christmas has been around for far more millennia than even the Abrahamic religions, and this humanist and her family have always adored it.

We put up as big a Christmas tree as possible, and set it blazing with very modern fiber-optic lights, old-fashioned tinsel, baubles galore, and an outrageously huge pile of mysteriously wrapped parcels beneath. We're a big friendly family and tend toward lots of people on Christmas Day. We've had as many as thirty; the reason for this will become apparent in due course.

We light scented candles everywhere in place of bonfires and to decorate the house use lots of holly (lucky us—a bush in the garden) and ivy, chosen as a symbol for the festival by those ancient Norse people who had discovered, as modern gardeners do, that it is the most stubborn of plants and however hard you try to get rid of it, back it comes every time.

Finally, a socking great wreath of holly, ivy, red ribbon, and tinsel is put on the front door to encourage passers-by, and we settle down to the overeating, some overdrinking (not fond of that myself), and the blissful swapping of presents.

And in our case, very few arguments. I found a way that prevents it for our family, which I will now offer you. And there are a couple more ideas you may find useful.

ARGUMENTS

There are several reasons for arguments on Christmas Day. Some are obvious. Too much to drink is well up the list. Alcohol, contrary to popular belief, is a depressant, and too much too early in the day can cast guests and hosts into existential misery, which can be picked up by children, even babies, and make them fight with each other and bawl and generally make a very unfestive noise. Lack of exercise is another bad thing that can turn a temper sour. A joint walk by the whole crowd can be fun and very helpful if you're feeling stuffed. Or play vigorous games in the garden, if you have one. And there's a lot to be said for opening up family bedrooms and letting the older people snooze. They'll love you for it. There is also a special system of my own which works like a dream, as you will see later.

ALCOHOL

If your guests must have a drink as soon as they set foot in your house, then make sure it's not too potent. I opt always for bubbly; Champagne if we're flush, cava if not. Such nice people at Waitrose; they always have a vinous bargain or two at the end of the year.

And it is never offered plain. Guests can have it neat if they like, but in my experience offering freshly squeezed (no concentrates!) orange juice to make Buck's Fizz goes down well with almost everyone. Or try black currant syrup—a kir not so royale. Children given orange juice or black currant in the same glasses as the grown-ups feel no end of posh. No one gets unpleasantly drunk on Buck's Fizz—just agreeably merry. (Having something to nibble on with their bubbly also helps, by they way. Not too much, though; ruins their appetites for the big feast.)

Now that children have been mentioned, a necessary warning: however late it is when all the guests have gone, and whatever mess you leave to deal with in the morning, always collect all the bottles and glasses, especially half-empty ones. Children can poison themselves by drinking the dregs when they come down from their beds while you sleep blissfully on Boxing Day morning. Small amounts of alcohol are strong poison in small bodies.

BIG FAMILY FIGHTS

And now, the biggest discovery I ever made about avoiding unpleasant confrontations of the sort that can, at their worst, lead to family breakups with siblings refusing to speak to each other ever again and in-laws storming out of your house forever, even before the turkey gets out of the oven.

If you have relations you absolutely loathe and always have, and there is no hope of reconciliation, and if failure to invite them to all your festive events sets up a hissing and a gnashing among your nice relations as well as some friends, there is only one answer.

Treat them like grains of sand in an oyster. Grains of sand so domiciled will be covered by the highly irritated oyster with layer after layer of (I must be honest here) somewhat snotty goo. It may take a long time and lots of layers, but in the end, there it is. A beautiful, opalescent, lustrous, adorable pearl with a grain of sand deep in its heart.

The first year I tried it, inviting as many people as I could seat at the table, adding card tables and such to the dining table proper, fifteen or so in all, I didn't have to speak to the unwanted guests at all. Just a brief "hello" and "goodbye" and the day had been wonderful. Even the work wasn't too hard. It didn't cost a great deal more in viands; it did in vino, of course. But it's the festive time! And cooking for twenty-five creates the same amount of labor as cooking for ten, truly, and not so many leftovers. I've been doing it for years.

When we had to move because the kids kept growing and the walls started to bulge, I increased the number of guests I invited and included a couple more unlikeables. Great pearly parties, every time.

OVERSPENDING

Frankly, I refuse to discuss this at all. What right have I to tell other people how they should use their money? If they choose to go into debt to have a glorious Christmas, that is their affair. If they have lost their jobs and want to spend their redundancy money on just one day of hip hip hooray and buy their children hundred-quid presents, then that is their business and not mine. I do resent the gloomies of this world telling more cheerful people how to live their lives and complaining about our spending too much and forgetting the Deep Meaning of the time.

Pooh to Deep Meanings as far as I am concerned. All any of us have is the here and now, the lives we are living at the moment. We cannot know when we will die, and as a humanist, I know my universe will die with me and there'll be nothing to do and no one to have any fun with. So, eat, drink, and be merry. And have a very humanist pagan Yuletide.

I'm Dreaming of a Green Christmas

Siân Berry

Being green and being religious can fit together well. Most religious texts contain advice on caring for natural things, and most religions use these teachings in campaigns to cut carbon and save the planet ("What would Jesus drive?" being my particular favorite). I applaud all of this, but my own political and environmental activism comes more or less directly out of my non-belief in God.

I have always been an atheist. At Sunday school I was willing to learn the moral lessons of the parables but very unwilling to believe there was anything supernatural about the commonsense advice Jesus was doling out. Later, even under extreme peer pressure when most of my friends became Baptists in the middle of my teenage years (and despite the exciting prospect of a dramatic full-immersion baptism ceremony), the idea of a divine being was something I just couldn't accept.

I wasn't political either. At university, I studied metallurgy and the history of science, visited coal mines and nuclear power stations with relish, and steered well clear of anything that smelled of ideology. On the college committee I was entertainments rep, keeping my fellow students amused with seventies discos and pub quizzes and not giving a thought to the future or the bigger picture.

It was only later that I became more thoughtful and developed a proper humanist philosophy. After college, I moved into a house with five friends who brought with them a massive collection of books, including piles of politics, literature, and history. So, with these resources at my disposal, I sat down to work out what I thought about the world. After two years of serious study, the conclusion was something of a counterpoint to the Atheist Bus Campaign's famous slogan. My version goes something like this: "Our planet, its civilizations, and its people are unusual and fragile things. There's probably no God, so we'd better look after them well."

Once I had this sorted out, I was ready to go. I spotted the Green Party, realized I agreed with most of what they had to say, joined up, and volunteered to help. And in the busy eight years since that day, I hope I have helped to make a difference to the way some people think about how to help the world get by. Meanwhile, I've spent most of my professional life working as a writer and campaigner and, in the process, learned an awful lot about how to fail badly at convincing people into a greener, more responsible way of life.

Which brings me to Christmas—a time of year with plenty of communications mantraps for both greens and atheists. These days, it's a Christian holiday in name only for most of us, and most believers would probably agree that it's gotten well beyond everyone's control.

What started out as a few days of festivities now lasts about nine weeks and seems to involve about a quarter of a million different acts of marking the occasion. And it's impossible not to take part because everything to do with the Christmas season, no matter how newly invented, becomes instantly "traditional." Secret Santa presents in the office, Harry Potter films at the movies, *EastEnders* on TV, chocolate fountains with the Boxing Day buffet. All suddenly compulsory, as if they had been around forever. And what about those prawn rings that always appear in seasonal supermarket adverts? Since when were frozen tropical crustaceans a staple part of midwinter cuisine?

And there I go, moaning like a big Scrooge. But, believe it or not, I do enjoy a lot of things about Christmas. It's the only time of the year where my voicemail and inbox calm down and I can spend a few days lounging around with my sisters without a bulging to-do list nagging at the back of my mind. At its simplest and most secular—as a family get-together to mark the end of the year—Christmas can be a joy. But it's so easy to let things get out of hand during the run-up and get swept away in a consumer frenzy that has a terrible effect on the planet. I would love to be able to reclaim this essential midwinter break from religions on one side and from commercial interests who have turned it into a festival of waste on the other, but it's very hard to do this without sounding like you want to spoil everyone's fun.

In communications terms, it's virtually impossible to strike the right balance when suggesting things to change. Those wanting to make the celebrations secular fall regularly into the trap of rash pronouncements against Christian words and imagery that are pounced on by the press, leading to headlines such as "Bureaucrats Ban Christmas Trees," "School Staff Ban Christmas Cards," or "PC Firms Ban Christmas Glitz" (these are all real headlines). Similarly, Greens, in seeking to point out the worst bits of wastefulness, risk sounding like out-and-out party poopers who would like to see twinkly lights prohibited and everyone's presents wrapped in old newspaper. We put the Green press office on "bah-humbug watch" every year, but even so, it's very hard to prevent this kind of impression from creeping out.

Another problem is that it's just so easy to resort to sarcasm and grumpiness when the consumerism of the modern Christmas is being rammed down our throats at every advert break. And on this point, we probably have staunch allies in most Christians, who would much rather celebrate the earthly appearance of their deity in a simpler, more puritan fashion. However, while they might come across as a bit gloomy if they moan about waste and avarice, at least they have a romantic alternative to offer, with a great story, classic songs, candlelit carols, and midnight masses.

Without a religious hook, secular Greens have more of a dilemma, because we haven't yet developed a way of talking about the need for change that is half as satisfying as simply bah-humbugging and slagging things off.

There are, perhaps, some steps we can take in the right direction. In theory, it shouldn't be that hard to synthesize secularism and respect for nature into a set of activities that people want to take part in. Early Christians did, after all, subsume a riotous and decadent pagan festival to create Christmas in the first place, and most aspects of ancient festivities that survive in traditional Christmas activities are those that relate not to pagan gods and beliefs but to the marking of the passing of the year. So perhaps we could revive some of these ancient season-based traditions while cutting down on the consumerism? That would surely be a sellable combination, if we can just avoid sounding like hippies.

On the other hand, calming things down might be an easier push. After all, by the time December comes around, what most of us need is a bit of a rest, and there's nothing less harmful to the planet than a nice sit-down. In fact, once the shopping is done and the family is digesting the dinner, our natural instincts on December 25 are pretty much carbon neutral. "Let's turn off the TV and play a game" and "Let's all go for a nice walk" are both festive suggestions with a carbon impact of approximately zero.

Much more exciting, and by far the best modern subversion of Christmas-style über-consumerism I have seen, is the multitalented performer, activist, and 2009 Green candidate for Mayor of New York, the Reverend Bill Talen. Reverend Billy is the parodic preacher famous for leading the Church of Stop Shopping in a crusade against chain stores in New York. His actions have inspired and amused the city for years and helped stem the flow of Starbucks, Walmart, and Disney into New York's historic and unique neighborhoods, while boring and alienating no one.

However, we can't all pull off a piece of mock-revivalist street theater when we are faced with questions of what to say and do about Christmas. At home, we need to be able to push things in the right direction in ways that require less chutzpah and far less talent. Most of all, we need some actual, on-the-ground practical suggestions.

Over the years, I've been asked to think about this question for numerous articles, books and—hardest of all—radio phone-ins, and have identified a few ideas for making the holiday less stressful and more ecofriendly. So here, with my bah-humbug detector turned up to eleven, are four perfectly reasonable ways to make the holidays greener and more constructive, without ruining it for everyone.

REPLACE CASH AND CARBON WITH THOUGHT

We all know that with gifts it really is the thought that counts, but we still end up stuffing shopping bags with irrelevant gadgets for the people we love. As well as these being pointless and forgettable gifts, we also risk annoying our friends and family by leaving them with cupboards full of tat and the problem of how on earth you recycle a fiber-optic golf ball polisher.

The easiest way around this, especially for people we don't know very well, is to rule out desk toys, juicers, and other attic fodder and go for non-material gifts instead. Vouchers for meals, downloads, books, and theater tickets should come in handy for everyone one day, and won't take up storage space in the meantime. Similarly, memberships for organizations like the National Trust or the RSPB gets your giftee an easily recyclable magazine several times a year, plus discounts and free entry on days out.

For closer friends and relatives, nothing beats the tribute of a well-chosen book or a sentimental souvenir of something you both shared.

My Christmas list always starts out filled with good ideas for these kinds of presents, and a few years ago, with local shops and secondhand booksellers being wiped off the high street, I might have had an excuse for giving up on finding what I want and getting my dad a golf accessory, but now that we have the power of the Internet, almost anything is possible. Want to find the program from that Cup final your uncle never stops talking about? Fancy a vintage Biba bangle for your mom? How about an early edition of the *Eagle* for your granddad, or a signed copy of John Lydon's autobiography for your brother? All virtually instantly tracked down with a few clicks and a credit card.

And the good green news is that not only are you recycling in a very chic way, but most Internet sellers of secondhand and vintage goods are also small, independent (though not necessarily local) businesses, maintaining their specialist status online and overcoming the challenges faced by retailers who rely solely on footfall in shops.

"But what about the transport miles?" I imagine I can hear you say. Well, while it's not the same as popping out to an "emporium of everything" on your street corner, Internet purchases aren't as wasteful as you might think. By relying mainly on the regular mail service, the extra impact of small items is tiny, and even for large items, the savings compared with collecting the goods yourself are substantial. Of course, delivery vehicles still have to take to the roads, but they can visit many different houses in one trip, so the total number of journeys is reduced. Studies across Europe have shown that where people generally get to the shops by car, shopping from home results in more than a 70 percent reduction in the traffic miles involved in getting the same amount of goods to people's homes.

THROW A PARTY

I'm against guilt. I want to change the world by making green options the most obvious, cheapest, and above all easiest things to do. Since I started out in politics, my mission has been to take the difficulty out of being green and make it second nature and, ideally, invisible.

So, for me, the best green measures are done secretly or under cover of a completely different purpose. When you give people free loft insulation, ask them to thank you not for cutting their carbon footprint but for cutting their heating bills. Similarly, when you replace your town's regular Christmas decorations with strings of LEDs, point out how much prettier they are rather than bore people with how much less energy they use.

My second tip is therefore to host a Christmas party and make it secretly green. You'll be able to rub your hands in carbon-miser glee at diverting people from ready meals or nightclubs for an evening, and you'll also give them a great time in the process.

There are lots of easy ways to quietly make a party greener, all of which would sound like the worst kind of nagging if you suggested them to someone who was inviting you round to their house, but with you in charge, you can go to town. Get local food, bulk-buy organic wine and local beer, use real glasses and plates instead of throwaway plastic (you can usually borrow spare glasses from the people supplying your wine), and you've virtually halved the carbon footprint of the evening already. Finally, track down a local firm with hybrid electric taxis and write it up with the simple word "taxi" next to the telephone. Almost everyone will use the number, and almost everyone won't notice that their strangely quiet journey home is also low-carbon.

GO GOURMET

A typical festive dinner can contain ingredients that have been transported over 30,000 miles, but it's easy to cut this right down, even in a supermarket, simply by picking native products off the shelf instead of far-flung alternatives: hazelnuts rather than brazil nuts, English beer rather than Australian wine, local ham instead of Indonesian prawns for the Boxing Day buffet.

But you can, of course, do even more. The original midwinter festival involved a feast of seasonal produce, embellished with preserved items from earlier in the year, so root vegetables, cabbages, sprouts, dried fruit, nuts, local cheeses, and chutneys are all real, traditional low-carbon fare. They are also easy to find in the seasonal markets that spring up everywhere in December. These markets are usually put together by farmers in your area, so this is also a good green way to boost your local economy.

Sourcing a non-imported organic, free-range turkey, duck, or goose for dinner isn't that difficult but is definitely more expensive. So choose a smaller bird to make up for the extra cost per kilogram and you'll get a non-frightened, tastier meal and fewer leftover bits to deal with afterward.

And you can cut down dramatically on food miles, packaging, preservatives, and leftovers by spending the break doing a bit more proper cooking. It's the one time of year when you might actually have the time to follow recipes from Nigel and Nigella, so channel your inner domestic goddess and get the kids involved in creating homemade biscuits and cakes or boiling up chutneys and jams to see you through the rest of the year in home-cooked gourmet style.

START A GREEN TRADITION

There's no escaping the fact that December comes around every year. Still, coming up with new strategies every twelve months for avoiding "traditional" rituals (both religious and consumerist) soon becomes very tedious, so why not start your own personal secular traditions for marking the middle of winter and, while you're at it, make these green activities too?

Some of the most heartwarming things about family holidays at home follow these lines already. Whether it's a game of Monopoly after dinner, a sing-along with Granddad playing the spoons, or dusting off primary school tree decorations made from pasta, these are some of the family rituals that, for me, make Christmas something worth taking part in, rather than being cynical about.

If you are now in charge of a family yourself, take advantage of your power and impose some homespun green traditions on your own kids. Perhaps you can help instill an interest in nature by spicing up a family walk with an annual competition to find and photograph animal footprints in the mud (or snow, if you're really lucky). Or make decorating the house more ecofriendly by setting aside a Saturday for making painted paper chains in a child labor factory disguised as a "fun art project."

The tree has plenty of potential for making a secular Christmas greener. A symbol of nature devoid of religious meaning, it's also ripe for adding your own traditions and having its impact on the planet reduced. The question of which is the best choice—artificial or natural—is one of the classic green arguments, with commercial interests on both sides putting out press releases claiming theirs is the greener option in the long run.

Personally, I'm not convinced by the claim that artificial trees are best, partly because I like the smell of a real tree so much, and partly because metal, tinsel, and plastic still release plenty of nasties as they are manufactured (unlike a growing tree). They are also virtually impossible to recycle once they wear out, and I don't believe they are reused as many times as the manufacturers claim.

So, on balance, a well-managed tree farm—ideally not too far from your house—is the better option. Try to find one that allows you to plant a replacement tree as part of a day out—something kids will love. And remember that each year 6 million conifers end up in landfills, so when you have finished with your felled tree, make sure to use your local tree recycling service.

If you have the space outside, you can do even better by using a growing tree that lives in your garden most of the year and is brought inside for a couple of weeks in December. This has lots of advantages, including seeing your tree grow up with your family, and a huge reduction in needles trapped in the carpet pile.

AND FINALLY . . .

Remember, whatever advice you follow, don't let your efforts get in the way of having a very merry Christmas!

How to Stop Worrying and Enjoy Christmas

Mitch Benn

I've always been a huge fan of Christmas. In most of my early childhood memories I'm either ecstatic 'cos it's Christmas, beside myself with excitement because it's nearly Christmas, or glum because it won't be Christmas for *ages*. Now that I'm a dad I love Christmas more than ever. I love Christmas from the bottom of my godless atheist straight-through-Dawkins-and-out-the-other-side heathen heart. Love it.

And yet . . .

Every year when Christmas hoves into view I find myself wading through another slew of articles and op-ed pieces about how I'm wrong to enjoy Christmas, how my love of the season marks me out as foolish or shallow, how my ability to take simple joy in simple things means I'm, well, simple.

On the one hand, proselytizers of every stamp tell us through pursed lips that what we call "Christmas" is a lie, a betrayal and bastardization of its "true" self. Meanwhile, the hipsters and style Nazis wax scornful on the naffness of Christmas and the patent rubbishness of anyone suckered in by such a woeful enterprise in crapness (they like their "nesses," those types).

This should, by all that's right and proper, be a completely redundant article. Of all the days in the year, the day no one should need to be told how to enjoy is Christmas. Least of all unbelievers like us.

Most of the year round we're the lucky ones in this respect—those of us fortunate enough not to be in thrall to ancient scriptures are generally freer to, in the words of the bus advertisement, "stop worrying and enjoy our lives." But Christmas is different, marked as it is with that whopping great "Christ" taking up the first two-thirds of its name. As the fearless rejecters of religion, shouldn't we be rejecting Christmas as well?

Nope.

Not at all. It's okay, really it is. I'll explain.

Of course, before I start, I must admit that I may have this easier than many of you, as I was what I once heard Jonathan Miller call a "cradle atheist." I was raised by non-religious parents in a non-religious household, I was—and remain—unchristened and unbaptized into any faith, and my attitude to religion has always been one of casual disinterest tinged with amusement and bemusement.

So I have never "rejected" religion, as it was never in my life to reject. I've never had to take a stand, nail my colors to the atheist mast, put my money where my mouth is, or resort to any other defiant cliché. I can see how someone "raised in the faith" who subsequently turned against that faith might feel obliged to purge his or her life of all aspects of observance, and how Christmas might be one of the things that needed purging. But in our house, while we didn't do God, we did do Christmas, and that's how it is in my own house now. Some might perceive an insoluble contradiction in this, but there really isn't, or at least there needn't be.

What it all comes down to is a question: what *is* Christmas? And the answer—for all of us, believer or otherwise—is that Christmas is whatever you want it to be.

You see, Christmas, like all living things, is evolving. It's been through many phases and guises and it'll go through many more. Given that the "eat, drink, and be merry" aspect of Christmas predates the "O Come, All Ye Faithful" bit by a considerable margin, and could even be said to take precedence over it, what then is the "true" meaning?

The answer, again, is whichever you prefer. Those of you who wish to restrict your participation in Christmas to reverent, even solemn observance of the rites and customs pertaining to the day in your particular faith, knock yourself out. Those of us who choose to celebrate Christmas in the traditional, pre-Christian manner (at least as it manifests itself in the modern era—basically eating forty mince pies and then slipping into unconsciousness on the sofa) may do so with a clear conscience. Spiritually, anyway. Nutritionally, that's another matter.

But shouldn't we, if we're going to ignore the "Christ" part of Christmas, find something else to call it?

This has been a divisive question in recent years, with right-wing commentators denouncing a pernicious, surreptitious, and largely imaginary (as these things almost inevitably are) assault by the forces of political correctness (boo!) on good ol' Christmas in particular and on Christian values in general.

Leaving aside the fact that nearly all the incidents of which this political correctness was supposed to consist turned out to be vastly exaggerated, if not completely made up (Google the word *Winterval* for a choice example of this), all of this is pointless divisive nonsense and entirely unnecessary because . . .

It's just a name.

That's all. It's just a name. *Christmas* is the name most of the English-speaking world currently gives the winter solstice festival, and it's as good a name for it as any other it's ever had. Yes, it's the name given to it by the Christian church in order to pay reverence to their figurehead Jesus Christ, as evinced by the aforementioned whopping great

Christ at the beginning, but so what? It's just a name, and the name of a thing can't be held to determine its form and function for all time, whatever the "reason for the season" crowd might claim.

Not convinced? Okay then, what day is it today? I have no idea what day it is where you are, but I'm typing this on a Wednesday. As such, I, in common with all respecters of tradition, have dedicated today to the glory of Odin. Man, I love Wednesdays. Bit of oar-running, bit of pillage, bit of—well, we'll crack open a bottle of mead and see where the evening takes us, shall we?

Maybe you're reading this on a Friday, in which case I do hope you remembered to honor the goddess Freya this morning. Seriously, you know how she gets.

This book is due out for the Christmas market, so perhaps you're reading it in November or December, which are, obviously, the ninth and tenth months of the year. And once Christmas is out of the way I'm certain you'll all be sure to see in the New Year by making sacrificial offerings to the Roman god Janus.

You see where I'm going with this? Christmas is named—as is just about everything else in the calendar—after a religious rite or holy day, but the mere fact that, unlike the Norse and Roman gods, Jesus Christ is still worshiped by some people, doesn't make the name any more "sacred" than that of any other day.

Moreover, if anyone tells you that the name "Christmas" renders December 25 a uniquely and quintessentially Christian affair, then you might ask him what he's going to call Easter from now on, 'cos if the name of a festival dictates its nature, then Easter—named for the goddess Oestre—has nowt whatsoever to do with the crucifixion or resurrection of anybody, and is still the frolicsome spring equinox fertility rite it always used to be. (Yes, it is; what do you think eggs and bunny rabbits are supposed to symbolize?) If the person in question is American, you might further inquire as to whom he considers to be the founder of his nation. Deduct a point for every name he suggests before he gets around to the Renaissance Italian cartographer Amerigo Vespucci.

(If you're still in the mood, ask him why people in New York don't drink bitter, wear flat caps, and say "'appen" instead of "yes." You'll have made your point by now and just be annoying him, but what the heck.)

This has all become a bit baroque, so by way of summary let me say this: if only practicing Christians can use the word "Christmas," then only Vikings can use the word "Thursday."

Of course, if you'd rather *not* call it Christmas, either because your own culture has a name for the season that you feel should take precedence (in which case, happy Hanukkah and thanks for reading this far) or because you just plain don't want to, that's fine too. But if you'd like to call it Christmas, either for the sake of cultural cohesion (still a good thing even without God at its center) or just for ease and convenience, then it's okay to do so. You're not trespassing on anybody's exclusive territory. It's fine. Go ahead. Say it. Christmas. There. That wasn't so bad, was it? Christmas. It's easy. Christmas. It's not even pronounced "Christ's mass" any more, it's "krissmuss." If anything, it sounds like it's been named in honor of somebody called Chris. Hark the angelic host proclaim, Chris is born in Bethlehem. Aw, how nice. Must send Chris' mom and dad something.

All good, insofar as it goes, but even if we're going to use the word "Christmas," surely we should forswear all the churchy rigmarole that goes with it, the carol services, the Nativity plays, and all that. Again, if you want to, fine, but if you don't want to—if your little darling gets cast as Mary and you don't feel like ruining her whole year by droning on and on about the historical inconsistencies in the gospels and the scientific implausibility of virgin births, if you find yourself in the mood to sneak into church on Christmas Eve and give "God Rest Ye . . ." some welly—this is all perfectly okay and in *no way* a betrayal of your Deeply Held Atheist Principles. Mainly because there's no such *thing* as Deeply Held Atheist Principles—it's not a belief system, it's the absence of one—but also because the Christmas story, like all the great legends, is still a rich and meaningful tale even if you don't believe any of it actually happened.

Just about the only thing I loved anywhere near as much as Christmas when I was a kid was *Star Wars*. I think I learned a great deal about sacrifice, strength, and perseverance from the tale of Luke Skywalker's quest to face and ultimately redeem his fallen father, and the Force is as vivid an analogy for the human spirit's capacity for good and evil as I've ever encountered, but I never thought it was, like, true or anything.

Similarly, the tale of Jesus is deep, fascinating, and moving; God Himself takes human form and lives a whole life among the mortals before being betrayed and murdered by His own creations in an attempt to offer the supreme moral example. It's ripping stuff, let's face it. Frank Herbert couldn't have done better. The fact that little if any of it is based on historical fact is neither here nor there, and to pretend that this story hasn't played a massive part in the evolution of our civilization is disingenuous to say the least, even if we as freethinkers recognize that human morality is not derived from religious ethics (it's the other way round).

In any case, let's be honest—to what extent is Christmas actually about Christ anyway? We've developed a predominantly secular, post-Christian iconography for the festival . . . for every Baby Jesus one sees at this time of year, there are ten snowmen, or Christmas trees, or Santas (who may be supposed to be a saint but is a fairly pagan figure in appearance and demeanor, not to mention the fact that he's at least partly a Coca-Cola commercial).

We've even—since 1843 at least—acquired our own post-Christian Christmas myth. Dickens' *A Christmas Carol* acknowledges Christmas' religious foundations and has a healthy dose of the supernatural (and ends with the most famous "God bless us" in literary history), but it's a very human tale of the redemptive power of a festival celebrating all that makes life—*this* life—worth living. And it's this story that is told and retold every year, with literally dozens of film adaptations littering the festive TV schedules (there's a particularly underrated version starring George C. Scott I'm always looking out for), while the supposed *actual* Christmas story, the Nativity, is confined to primary school assembly halls and that Frankie Goes to Hollywood video.

So—celebrate Christmas *if* you want to, *how* you want to, and call it what you like while you're doing it. It's as much ours as it is anybody else's, and as much everybody else's as it is ours. And so, as Tiny Tim would have observed if he'd grown up in my house, random circumstance and the smooth operation of the laws of the universe bless us, every one!

How to Decorate the Outside of Your House with Lights and Not Have Your Neighbors Hate You:

A Guide to Turning Your Home into a Festive Something That Is So Bright It Can Be Seen from Space

JON HOLMES

This year, as you sit in the gap between Christmas and New Year, idly crushing liqueur chocolates into your mouth, wondering why each and every one tastes the same no matter what liqueur it purports to contain, what will you do to pass the time? Perhaps you'll get round to watching that DVD box set of television you got ("All of television! On a billion disks!"), maybe you'll suggest playing traditional yet baffling games ("Who's for Barrymore Cluedo?"), or you may be considering combining the two and playing games while watching television by switching it on and then trying to guess the number of sweets in Eamonn Holmes.

But wait. May I politely suggest an alternative? This year, instead of lying fatly on the sofa under a flimsy cracker hat, by now spooning pickle into your gob because you've run out of chocolates, why not simply sit quietly at this special time of year and take some time to reflect upon all of mankind's greatest achievements? Sit and marvel at everything science has achieved. What we've learned, what we've built; how far we've come medically, creatively, and technologically. And then, when you've done that, try to figure out just how, in the several billion years since we flopped out of the sea and grew legs, we appear to have reached a point where someone considers it acceptable to have taken all of our knowledge of light and electricity and used it to fashion a six-foot-wide snow globe with a festive scene in it that's specially designed to sit in someone's front garden each year looking inexplicably cheerless.

Well done, science. You must be very proud. How did that conversation in the laboratory go?

"What shall we do today? Invent a time-travel device, find a cure for cancer, or come up with some sort of animated globule with a Santa in it that will look stupid outside people's houses at Christmas?"

"Ooh, ooh, the third one! I'll fetch the tinsel!"

I have a strange relationship with outdoor decorations. I don't mind a tasteful Christmas light or two. I'd happily slow down to look at a small white bulb glowing here or a splash of illuminated tree there, but the ones I'm talking about are those that cling on to the front of houses like flashing, electricity-sucking ticks.

You know the ones I mean. You're driving along of an evening, and then suddenly, as you round a corner, it's like someone's inadvertently opened the gates of hell during Satan's office Christmas party, and before you know it, all the garb from Dante's tenth, lesser-known but equally unpleasant circle (the Walled City with Kitschy Crap All Over the Wall) has burst out and become stuck fast to the brickwork of a drab semi.

There's a village just down the road from where I live that, at this time of year, can be seen from space. It's as if the residents have had a competition to see whose house can be made the most hazardous to passing air traffic. Honestly, it's like everyone who lives there has spent the last few years rushing down to the remains of Woolworth's, gathering up armfuls of anything that glitters or lights up, rammed it all down the endpipe of a shoulder-mounted missile launcher, and then fired the whole lot at their dwellings like a gaudy bomb, a bomb that's spread out on impact and covered their walls and roofs in a furious splatter of garish Christmas trinkets.

Now, if you've done this, I'm sorry to have to tell you that chances are any of your neighbors who haven't done this will want to kill you. And not quickly either. They probably want to take the tangled wire from that shitty angel that you've got in a tree and hang you with it in the center of the road as a warning to others. And then wrap your body up in the tattered remnants of a punctured inflatable Father Christmas and leave it on the doorstep of Argos to dissuade anyone else from ever buying one.

So the question is: what do you do if you want to seasonally decorate your house without everyone around you wishing both you and your electricity supply dead? Well, fortunately for the residents of the aforementioned village, everyone who lives there joins in, so as you're kept awake between the hours of November and January by a glow not dissimilar to a melting reactor pressing against the curtains, you can rest assured that your 8-billion-candlepower neon Santa on the chimney is burning your neighbor's sleepless eyes off as well. And this may well be the answer. Nope, you can't go it alone with a glowhome or your neighbors will simply spend Christmas hoping for an electrical fault that burns your house and all its adornments to bits. The only way forward is to get everyone else to join in. It's all or nothing in the world of animatronic snowmen and sequencing rope lights, because if it's just you who's got a life-size incandescent Nativity scene on the lawn and no one else, then you'll stand out like a moob at a gym and everyone will hate you.

Actually, while we're on the subject, there's another thing. Why are people no longer satisfied with a few simple reindeer or a straightforward wall-mounted flashing "Merry Xmas" sign? For some reason, there are now various lit-up depictions of Santa using modes of transport with which he's not normally associated and which, quite frankly, would be a hazard to any form of rooftop present delivery. Sure, there are the usual smattering of sleighs, but now, every year, from the end of October onward (when somebody, presumably a git, has decided that Christmas begins), there are any number of lit-up representations of Santa riding everything from a hot-air balloon to a train and a bike. In exactly what version of the story of St. Nicholas did he bring presents on a bike? In no version, that's what version. Is it because a bike or balloon is a modern, greener alternative to a mystical sledge? Or because they're cheaper to run, what with the spiraling cost of reindeer feed in a global downturn? Whatever the reason, it plainly makes no practical sense, because whichever way you ride it, a train, bike, or balloon just wouldn't be able to cope with the rigorous demands of traveling round the world in one night, stopping off to deliver presents.

For a start, storage on a bike is notoriously poor, and don't even get me started on the reliability of a train. Santa would optimistically board the West Coast Mainline with his sack of presents, and then only get as far as Nuneaton, where he'd have to pick up the replacement sled service because of planned engineering works at the North Pole.

Another "decoration" I've seen was bolted to a bungalow and had Father Christmas going up and down a tasteless, flashing ladder. Duh. Santa doesn't need a ladder, does he? For one thing, he's magic, and for another—it's a bungalow. All Santa would need to do is stand on the wheelie bin and use a drainpipe for purchase. There's certainly little point in his carting a bloody great ladder around. It's completely unrealistic.

So anyway, if, to you, Christmas wouldn't be Christmas and would simply be ruined if your house wasn't radioactive, then here's an idea. Let's suppose that I'm right and the only way to have outdoor lights and not be killed is to make them communal. All you need to do is make the community bigger. Global, even. If you're going to blind people with Christmas, then you have to make it count. So here's my plan: why don't we just make one great big simple glowing outdoor decoration that we can all hate together as one? Why not take every meretricious outdoor Christmas light on the planet . . . and put them on the moon? Let's turn the moon into a giant astrophysical bauble—plus it's got a head start because it glows at night already.

And don't worry about the logistics of having to collect up all the decorations on Earth and ship them to NASA in readiness for Operation ChavPlanet, because I've already thought of that too. Gathering up all the flickering angels and Santas and mangers and donkeys and snowflakes would be easy because all you'd have to do is tie one end of a rope to the plastic Rudolph next door and the other to the tail of the space shuttle while it was on the launchpad. Then, when it blasts off and leaves Earth, it would rip all the outdoor decorations out of all the gardens of the world in one go and yank them all into orbit like a giant space sleigh. And once they'd been dropped off on the moon, the newly astro-trained residents of the village near me could be standing by to decorate the lunar surface, while those of us left on Earth would stare humbly into the sky knowing that man is up there, attaching Santas on ladders to the sides of craters and rigging up a plastic waving fairy in the Sea of Tranquility.

And then, when it's switched on, as tiny beings on one small planet floating alone in the inky, majestic blackness of space, we could all hate it together. One small step for a man, one giant, gaudy waving snowman for mankind.

How to Escape from Christmas

ANDREW MUELLER

Some years ago, a magazine that retained me as a columnist and resident curmudgeon asked me, in the interests of balance, to dash off a few paragraphs of Scroogeish grumbling for their Christmas issue. It ended up amounting to 406 of the toughest words I've struggled to compose. It wasn't that I had trouble identifying aspects of Christmas that I found objectionable; these were, and are, giddyingly manifold. It was that enumerating and explaining them felt—even by the existentially excruciating standards of writing op-ed commentary—hackneyed and futile.

Every complaint about Christmas has been made, and made often. It would be utterly unsurprising if a hitherto undiscovered gospel, disinterred from some Dead Sea cave, recorded a previously undocumented witness to the Nativity of Jesus reacting to the arrival of the Magi, and the gifts of gold, myrrh, and frankincense they bore, with an indignant tut of "Tch, it's all got so commercial."

Nevertheless, I tried. I enacted the ritual rage against annoyingly cheerful television commercials, tacky shop windows, witless seasonal novelty pop hits, bumptious television presenters, stupid hats, shrieking idiots staggering out of office parties, the incessant instruction to Simply Have a Wonderful Christmastime, the annual exhumation of Slade and Wizzard ("I Wish It Could Be Christmas Everyday"—hardly surprising, given that it's the one day in any calendar anyone still gives a figgy pudding about Roy Wood). I harrumphed about the fact that the whole dreary circus begins earlier each year—by my admittedly unscientific calculations, it will be as soon as 2037 that the first Christmas decorations go up before Easter. And I moaned that the worst thing about feeling like this was knowing that the only people equally vexed at the reduction of the Great Redeemer's birthday to this grotesque and fatuous carnival of consumption were the sort who'd rather spend Christmas—and, indeed, every Sunday morning—kneeling on cold stone and mumbling in Latin. "Bah!" I concluded. "Humbug!"

It made no difference, of course. Observant readers will have detected that, even despite the most strenuous efforts of a commentator so widely respected and feared as yours truly, Christmas continues to be celebrated—and perhaps, I will bemusedly concede, even enjoyed—by substantial tranches of humanity.

In some respects, I have no problem with this. Though I suspect that many of the yarns spun about Jesus of Nazareth by his disciples have been somewhat tweaked for dramatic effect—and a few of his entourage were, after all, fishermen, a profession proverbial for a diffident relationship with the truth—I greatly admire what seems, in my admittedly inexpert view, to be the message at the core of Christ's preaching: i.e., try not to spend your brief time in this corporeal realm acting like a dickhead, and be mindful of the other chap's point of view if at all possible. There are worse historical figures for whom we could insist on throwing an annual planet-wide party. Like, for example, almost all of them.

So it's not the fact of Christmas but the manner of it that depresses and alienates: the vast coercive conspiracy to shrilly insist that you do and think absolutely the same thing at the same time as everyone else. So overwhelming is it that even those of us whose instincts would tend away from leaving brandy and biscuits out for Santa Claus and more toward acknowledging the night before Christmas by planting a bear trap in the hearth grudgingly discharge their festive duties: eating too much; drinking what proves to be never quite enough; spending money on, and time with, people we spend the rest of the year avoiding; slumping in front of our seventy-third viewing of *The Great Escape*; feeling a stultifying empathy with the cooler-bound Steve McQueen, and idly daydreaming of a tunnel under the wire, forged travel documents, and freedom.

Happily, there are options. The most prosaic of them is to spend Christmas Day at home, alone. This is more difficult than it sounds, and the reason why is the best illustration—and condemnation—of the oppressive groupthink that Christmas engenders: nobody wants to let you. When you announce your intentions to sit proceedings out, people—whether motivated by guileless charity, a faint hope that the presence of an outsider may serve to dilute their annual family bloodbath, or outraged resentment that someone is going to get away with it—will insist that you spend the day with them. It is therefore important that your preparations for a solitary Yuletide are conducted in secrecy, perhaps with the aid of the useful misdirection always available when one is invited to several things on the same date, i.e., telling anyone who asks that you've already taken up another option. If you pull this off, the rewards are considerable, and perhaps even appropriate: the most peaceful day you'll have all year. Catch up on reading. Go for a long walk—you will own the streets of even the most bustling conurbation. Revel in the thought of those less fortunate or determined than yourself, feigning excitement as they remove the wrapping from gifts they didn't want presented by people they can't stand. Enjoy a serenity unbroken by the flying crockery with which so many people's nearest and dearest express their joy at having this precious time together.

It may be pointed out that this course is hardly suitable for readers encumbered with children of their own, or similar signifiers of something dimly recognizable as a grown-up life, like a spouse or indeed any palpable responsibilities whatsoever. There are two possible responses to this. The first of these, "Ha ha ha ha ha ha," may be interpreted in some quarters as not constructive. This being the case, it's best to move on to practical escape plans that can be made, if necessary or desirable, in concert with others.

For anyone living amid any culture rooted in Judeo-Christianity, the very idea that Christmas can be escaped may seem absurd. For a quarter of every year, the music they hear, the periodicals they read, the television they watch, and the streets they walk are dominated by reminders of who is unchallengeably in charge—it's like spending 25 percent of your life living in a dictatorship, albeit one tyrannized by a fat bearded bloke in pajamas and boots. It is also, sadly, increasingly the case that even those parts of the world whose predominant faiths or ideologies theoretically preclude them from subscribing to this gaudy inanity—the 1.3 billion Chinese and 1.1 billion Indians who (mostly) shouldn't care, for a start—are buying into it anyway. That they are doing so for nakedly commercial motives is commendably honest, but it's no less annoying.

There is much to be said, therefore, for decamping to somewhere people believe some variation on the monotheistic legends. The people of the Islamic world revere Jesus of Nazareth as an all-round good egg, and are also keen on his mother—Mary is the only woman mentioned by name in the Koran. Muslims in Muslim countries do not, however, make an untoward carol and dance about Christ's birthday. I discovered this a few years ago, when I happened to be spending December in the Holy Land, the place where Christ's message was first propagated (and, of course, where it has since been most assiduously ignored). On Christmas Eve, in my hotel in East Jerusalem, just a block away from the Damascus Gate, I considered my options for the following day. The Old City of Jerusalem itself, I thought, would be a bit much (the Old City of Jerusalem is always a bit much). Going to Bethlehem, a taxi ride and a couple of Israel Defense Forces checkpoints away, might make me feel like a gatecrasher—for the people who'd made the effort to be there, being there would be a big deal, and I had no wish to intrude (for the same reason, though I enjoy visiting churches, temples, mosques, and synagogues when I'm traveling, I always keep away during services, unless invited). So I called my Palestinian fixer.

"Let's go to Hebron," she suggested.

Hebron is known for, among other things, being the burial place of Abraham. By some interpretations, the entire millennia-spanning gory travail of human history has been essentially a family feud among his descendants. Spending Christmas with the patriarch instantly struck me as either devastatingly apt or wildly inappropriate, and either way, an excellent suggestion.

And so it proved. As a writer with a straitened budget, I was relieved to discover, upon disembarking from the taxi that had borne us from Jerusalem, that this wouldn't be a dead day in my itinerary. Everything in Hebron was open, aside from the old city's shuttered and largely abandoned marketplace. This was closed not temporarily out of religious observance but permanently because it had become a front line in the internecine squabbling of the great-great-umpty-great-grandchildren of the chap entombed at the far end of it. But away from the parts of town annexed by Jewish settlers and the Israeli soldiers protecting them (rather reluctantly, more than one was willing to admit), Hebron had the amiable bustle typical of Arab cities. The shops, cafés, and restaurants I visited offered an added seasonal bonus: I was barely allowed to pay for anything.

"They're usually pretty friendly to visitors down here," explained my fixer over lunch, "but they know it's Christmas Day, and they think you're a Christian."

I sipped my complementary coffee, returned the smiles of the beaming restaurant staff, and felt vaguely guilty.

"Not," she continued, "that they'd care all that much if you told them the truth. They are genuinely honored."

I asked her to requite the sentiment on my behalf, my Arabic not being all it might. And I wondered, if one took a global view, how many people hosting other people were, right at that moment, regarding their guests with such pure-hearted generosity. If someone can conclusively demonstrate that the percentage clears double figures, I shall strap on a pair of antlers and guide Santa's sleigh personally.

Delightful though Christmas in Hebron was, it still felt a bit of a cheat, and I worry that any similar excursion would also induce the nagging sensation that one was benefiting from hospitality under slightly spurious pretenses. You're not really avoiding Christmas if you're spending it among people who are being nice to you because they assume you'd rather be celebrating it. Given that the same unease will afflict anyone deflected from the home-alone strategy by the kindness of friends, it is clear that more drastic measures are called for.

The invention of a time-travel device that would enable one to skip December 25—or, indeed, to hurtle backward roughly 2,000 years to the Middle East with a view to asking one well-meaning carpenter if he was certain he'd thought this whole thing through—is probably still some years away. The next best thing does exist, however: the combination of passenger jets and time zones. On several Christmas Eves, I've joined kindred souls at Heathrow, climbing aboard the flight that takes off late on December 24, refuels briefly in some conveniently unchristian way station—Dubai, Kuala Lumpur, Hong Kong, Bangkok—and arrives on the east coast of Australia early on Boxing Day. Not only do you miss the whole thing, you've access to the greatest jet-lag recovery cures available—sunshine, leftovers, and live coverage of the Test match.

Though it's unfair that one is driven to such lengths, such lengths are necessary. Because it's not just about avoiding Christmas. It's also about avoiding the assumptions and the jokes that people make about you when you do. These are largely the fault of Charles Dickens. It's thanks to his creation of Scrooge that anyone who voices the most reasonable of objections to Christmas is reflexively dismissed—or, worse, patronized—as a mean, tight-fisted, lonely ogre. Which is like believing that all vegetarians nurture ambitions of remilitarizing the Rhineland, invading Poland, and building a Reich that will last a thousand years.

Let the record show that I enjoy all the following: eating well, drinking a lot, spending time with friends and family, giving presents, receiving presents, getting cards and e-mails from people I haven't heard from for a while, and Phil Spector's Christmas album. I'd just prefer to enjoy all those things when I decide to. Declining to take part in Christmas isn't easy, but conscientious objection of this sort is liberating, and satisfying—and, for what it may be worth, much more respectful of Christian believers than trudging along with it. (Along the same lines, though I have in my time quite sincerely wished people in various locations all the best with Ramadan, Diwali, Ashura, Yom Kippur, or whatever they were having themselves, it would have felt ridiculous to celebrate any of these personally.)

I've never believed in God, but I've never wanted to prevent anyone else from doing so, if that's what gets them through the night. This life is short and this world is weird, and the appeal of an off-the-peg explanation for everything is not hard to discern. All I've ever asked of the faithful is that they keep it to themselves, and avoiding Christmas is my way of extending them that courtesy. I encourage my fellow unbelievers to consider the same course, and to book their flights. Christmas: if you can't join it, beat it.

Gifts for the Godless

Jennifer McCreight

If there's one Christmas tradition that both atheists and Christians can get behind, it's the presents. Sure, cynics on both sides may sneer, claiming that we're just buying into capitalistic, materialistic manipulation . . . and true, anything that can make people trample each other over a giggling Sesame Street character *should* terrify us all. But can we really ignore the delightful joy that trading presents brings?

Now, I probably don't have to explain why anyone under the age of eighteen loves getting presents. Even the most humble kid will be happy to play with a new teddy bear, dance to a new CD, or rip open a new pack of Pokémon cards, desperately looking for that Charizard (maybe that last one was just my generation). But even adults love exchanging gifts—and for all but the most profoundly selfish, it eventually becomes less about receiving presents and more about giving them. Despite that warm, fuzzy (and possibly optimistic) thought, there remains one serious problem with secular gift giving: deciding what gift to buy.

I know what you're thinking: *everyone* has a hard time figuring out what to buy a person! But do you know how difficult that is when the receiver is an atheist? It immediately rules out a whole slew of holiday gift options. No ornaments (they're hard to hang on a Festivus pole), no cross necklaces (they irritate the skin of the godless), no cheesy Christmas sweaters (hey, we still have *taste*). Book selection is limited—*Baby's First December 25* doesn't exactly have the same ring to it. And don't even *think* about using that Baby Jesus wrapping paper.

Now, this isn't to say you can't buy an atheist *any* religiously themed gifts. The main exception I can think of is a Nativity scene. One man's representation of the birth of his savior is another man's new set of action figures. A gift of the Bible may also work, as long as the margins are large enough for note taking. And who knows, maybe your godless buddy takes some ironic delight in collecting religious kitsch. One of the most hilarious Christmas gifts I've received was a box of over a hundred Chick tracts. (Thanks, Erin!)

Regardless of your religious belief, you can surely enjoy gifts like toys, MP3 players, and clothes (well, unless they're a wool-linen blend).* But where is all the explicitly atheist-friendly Christmas merchandise? There is a gaping hole in the market that someone should seriously be capitalizing upon. We can read only so many blasphemous books and wear only so much irreverent clothing. Personally, I would have loved for someone to give me some godless toys when I was younger (but not too young; let's leave indoctrination to the theists).

We've stumbled upon a new market, so I have some suggestions for merchandise:

1. *The Deity Birthday Match Game.* Educational games may not have lasting appeal, but this one can make a great gift if you're on a budget.

---

* Deuteronomy 22:11.

Everyone has played match games before: spread out the cards face down, flip two at a time, and keep them face up if you find a matching pair. Here, half of the cards would list deities from different religions, and the other half would list their birthdays. What better way to promote religious literacy than with this lightning-paced, difficult game of chance and memory? Jesus would match with December 25, Attis with December 25, Dionysus with December 25, Osiris with December 25, Mithra with December 25 . . . Hm, maybe this game would be a little too easy.

2. Designer-less Legos. If you're an atheist or even remotely interested in evolutionary biology, you've probably heard of the watchmaker analogy. Repeatedly. It's a teleological argument for the existence of God that basically says, "Something that looks designed must have a designer!" Creationists are so annoyingly persistent with this piece of intellectual laziness that it still pops up in apologetics, even though Richard Dawkins wrote a whole book refuting it almost twenty-five years ago.*

Why do we keep hearing this argument after it's been given a good thrashing by a number of scientists? One word: Legos! Think about it. Aren't we just priming children to think design implies a designer if we give them explicit instructions on how to build a space station or dinosaur out of little plastic blocks? Now, Legos were my favorite toy as a child, and I still became a godless evolutionary biologist. But I am merely one data point. Should we really be taking such a risk with our children?

There is an obvious solution. Instead of giving our children physical building blocks, we can give them a computer simulation. The simulation will start with hundreds of the same Lego block, and

* Richard Dawkins, *The Blind Watchmaker* (New York: W. W. Norton & Company, 1986).

random blocks will start adding themselves to the piece. Gradually the construction will grow more complex, with the most structurally sound construction propagating its pattern throughout the "population." Of course, to keep things fun, we don't have to make reproductive success based on structural soundness. I'm sure we can figure out how to code for other selection pressures kids would be interested in, like "awesomeness" and "maximum number of blinking lights."

Yes, this simulation would take quite a long time to run, and it wouldn't require any interaction with the user, since that's not how evolution works. But certainly we atheists can all agree that scientific accuracy must take place over having fun! Hell, we all saw what happened when the game Spore tried to go the fun route: it made biologists weep *and* was dull.

3. *Grayscale crayons.* To represent how atheists view a bleak world devoid of divine purpose and meaning.

4. *Atheist Barbie.* I've never been a fan of Barbie. Even though some family members frequently bought her for me in a desperate attempt to get me interested in something girly, Barbie and I just never clicked. Now, as a feminist, I have even more reasons to dislike the doll. But I know there are plenty of girls and boys out there who love Barbie, and who am I to deny them that choice?

So I have a better solution than an outright ban on the buxom beauty. She's already had every career from ballerina to doctor, from firefighter to computer engineer. It's time for a Barbie that atheist women can relate to!

Atheist Barbie comes with many different features that represent the everyday life of a female atheist:

1. Stylish nerd glasses
2. Laptop bag for easy access to atheist blogs and emergency use of Snopes
3. Flying Spaghetti Monster necklace
4. Geeky pro-science T-shirt
5. Lunch
6. Atheist pieces of flair (the Out Campaign's "Scarlet *A*," Darwin Fish, and Squid—Cthulhu and Invisible Pink Unicorn sold separately)
7. Godless reading material
8. No pants, to be ready for frequent surprise orgies

I'm not sure how you could ask for a better role model! All we'd need is homosecular gaytheist Ken to be by her side.

Maybe my atheist toy ideas need a little more time to evolve. But until we come up with something better, I guess we can spend a couple more Christmases buying each other books and bumper stickers.

PHILOSOPHY

I'm an atheist, and that's it. I believe there's nothing we can know except that we should be kind to one another and do what we can for other people.

—Katharine Hepburn

If God Existed, Would He Have a Sense of Humor?

Charlie Brooker

By any standards, God is a coolly uninvolved sort of character, content to sit back and watch as mankind has one bucket of peril after another tipped over its collective head. He witnesses deaths, disasters, wars, diseases, and the continued existence of Razorlight and doesn't lift a finger to help, except to whisper murderous instructions into the mind's ear of the occasional insane truck driver. If he's truly omnipotent yet refuses to intervene, there must be another reason. Such as laughter.

Phalaris, the mad tyrant of Acragas, who ruled Sicily from 570 to 554 BC, had a bloodcurdling torture device called the Brazen Bull built for his amusement. It was a hollow bronze bull into which miscreants were placed. Flames heated the bull from below; as it warmed up, its victims were cooked alive. Their agonized screams would travel from the enclosed belly into the bull's head, where a complex system of pipes and horns mutated each shriek into a comic mooing noise, which would be emitted through the mouth. Phalaris reportedly found this hilarious (although it probably struck him as slightly less funny when he was overthrown and tossed into the Brazen Bull himself).

Maybe God shares Phalaris' sense of humor. Perhaps he finds human suffering funny, much as I find it hilarious when Wile E. Coyote has his head stove in by a plummeting anvil in a *Road Runner* cartoon. Perhaps he's so detached, our lives are a mere cartoon to him.

After all, if God really is an all-powerful eternal deity, capable of observing the entirety of human history in the blink of an eye, he'd need a jet-black sense of humor just to stay sane. I get depressed after half an hour of Sky News, even on a slow day. If firefighters use gallows humor to defuse the tension after witnessing a tragic house fire, God would require a bulletproof comedic sensibility so sick it would appall Vlad the Impaler.

I'm not saying our human lives are an unending carnival of misery, incidentally. Far from it. We've got butterflies, for one thing. They're nice. Cookies are pretty good too. And we invented the amusement park. All I'm saying is that if God has half the powers attributed to him yet fails to lend a hand when things inevitably go wrong (which they do from time to time, because it's a big world: bigger even than Simon Cowell's garden), if God fails to prevent tsunamis and the like, then he must be getting something out of it. And that something must be laughter.

That would make him cruel. It's one thing for us mere mortals to laugh at a cartoon wolf being smashed on the noggin, one thing to laugh at a celebrity tumbling on the ice in a reality show, one thing to laugh even at a tasteless joke about a disaster or a famous criminal. But it's another thing entirely to laugh at something you have the power to prevent. That would be the laugh of a sadist, chuckling at a victim in a cage.

Don't know about you, but I don't relish the thought of a heartless, cackling, invincible boss watching over me, mocking my every setback. I've seen Gordon Ramsay's *Kitchen Nightmares*, and I'm thankful the real world isn't as horrible as that. I don't believe the boss is there at all. I think we're all freelancers. We're all in charge. Which means we should literally work together for the good of the company.

Preferably while laughing more than we currently do. Lift your head. Release the tension in those shoulders. And laugh. Because laughter's only human. Laughter keeps us in the moment and it keeps us on our toes. Laughter separates us from the gods while binding us closer together. If you're looking for a miracle, look no further than your most recent belly laugh. Maybe a friend made you clutch your sides till you shook with glee; maybe an old episode of *Frasier* had you howling on the carpet. Either way: in that moment, you were immortal. And that, my friend, is as sacred as it gets.

Unsilent Night

Hermione Eyre

Perhaps there should be some kind of course you can take in Atheist Assertiveness Training. I mean, there should definitely be a course you can take in Atheist Assertiveness Training. It would be run by someone rather in the mold of Barbara Woodhouse. "Louder," she would say as we shuffled up and down mumbling about how if it was all right with you we thought church and state should be separate. "And prouder!" she'd shout. "You'd be drowned out in seconds by a Sally Army band. You, in the tracksuit—say you don't believe like you mean it!" Then we'd hive off in pairs and practice non-aggressive confrontational skills before cooling off to a reading of Nietzsche's *Ecce Homo.*

Sure, it isn't going to happen, and that's partly what I love about atheism: the quiet individualism, the self-reliance, and the lack of enforced singing, organized sanctimony, and bake sales. But then again, it can get lonely out there in splendidly rational isolation. For the active minority, British Humanist Association lectures, campaigns, and meetings supply a sense of community among the godless. But in a general social sense, it's easy to feel alienated. Often, as with a minor ailment, you need to know someone quite well before you even realize they have atheism. You can begin a dinner party tirade without knowing if anyone round the table is going to back you up. At a church wedding it can be a surprise to see who else's head also refuses to bend in prayer; moreover, catching someone's eye during the Nicene Creed can imply many things besides religious skepticism.

It is hard to know how to—adopt American twang here, please—"self-identify" as an atheist. Fittingly for a once dangerous belief, it does not readily announce itself. Public burnings, dismissals from university, and so on have tended to disincentivize ostentatious displays of atheism over the years, and besides, freethinkers naturally mistrust uniforms. We don't want to conform to wearing our hair a certain way or adopting a vestigial hat. We have no symbolic trinkets, no Sikh *kara*s or Catholic rosaries; no arcane taboos, no dietary requirements, no cult pronouncements. How good it is to be free of these trappings; yet sometimes, without any rituals to observe or outward signs to flaunt, I find that my deeply held beliefs can feel insubstantial as air. I am not proud of this, but a small, atavistic part of me feels the lack of a badge or banner. These age-old urges die hard.

But to show our solidarity by wearing a regulation GOD IS DEAD T-shirt or an overpriced piece of string around the wrist would be wrong, wrong, wrong—and lazy. When you raise a totem you agree to be bound to whatever it symbolizes; you surrender independent thought. Better, surely, to express your atheism through a thousand rational acts; through constant low-level social vigilance; through countless tiny words and deeds.

As a movement, we rightly resist banding together—even writing "we" makes me feel a little coy—but we are never going to progress unless we're more vocal. I don't mean in the media, where atheism is well represented, but socially, randomly, impulsively.

It isn't easy being volubly atheist. We want to help spread the advantages of secular life, yet we don't want to evangelize. All too often these impulses cancel each other out and we end up doing or saying nothing. The non-believer has to learn how to pick a battle worth fighting—he or she has to discriminate between what is forthright, what is hectoring, and what is downright rude. It should be possible, after all, to be bravely and publicly atheist and still receive party invitations. Likewise, non-believers should be able to express themselves honestly and still have a happy family Christmas. Here are some worked examples of typically challenging situations:

1. A Jehovah's Witness rings your doorbell, saying, "Can I give you some literature about the Holy Bible?" Do you reply:
   (a) "Thanks, but I've got a bath running!"
   (b) "I don't believe in God. Can I give you some literature about humanism instead? Okay then, swaps?"
   (c) "Did they get you very young?"

2. While you are waiting to visit St. Peter's Basilica in Rome (just for the art, naturally), some devout children wearing mantillas push in front of you in the queue. Do you:
   (a) Say loudly, "Did you know some people don't believe in queue barging?"
   (b) Say loudly, "Did you know some people don't believe in God?"
   (c) Shout, "You need emancipating!" and try to remove their mantillas.

3. A friend's friend asks, in a voice that implies you are deeply shallow: "But do you really not believe in anything bigger than yourself? Do you have no spirituality at all?" Do you:

(a) Simperingly say, "No but I believe in, you know, art and people, and books, and nature . . . And I cried at the opera last week."
(b) Laugh and say, "Who are you, Angelina Jolie?"
(c) Say coldly, with slightly crossed eyes: "When you die you will rot and nothing of your ego will remain."

4. When a nurse remarks, "Maybe God will give her a baby!" about a mutual friend, do you:
(a) Laugh nervously, give up, and say, "Maybe!"
(b) Say, "Maybe she'll use contraception so she can decide when she's ready to give the baby the best start in life." (This is a point of view, after all, that a nurse ought to consider.)
(c) Snort and make the same point about contraception, followed by: "And it's wicked the pope doesn't let mothers in the developing world do the same." (Admirably bold though this is, she is a nurse, not an NGO.)

5. Your great-grandma asks you to drive her and your small cousins to midnight mass at Christmas. You are, apparently, the only person who is free and able. Do you:
(a) Concede graciously, take communion to please Granny, and keep your thoughts to yourself for once.
(b) Concede graciously, politely refuse communion but participate in the service with the calm, detached interest of an anthropologist, and then make some jolly conversation about the pagan roots of Christmas in the car.
(c) Drive them there scowlingly and spend the service sitting outside in the car listening to Kraftwerk and revving the engine.

You have now completed your Non-Aggressive Atheism Pop Quiz. Its purpose was, of course, not to moderate your convictions, simply their expression.

If you answered mainly (a): You need to crank it up, otherwise people are going to start mistaking you for an agnostic. Shudder.

If you answered mainly (c): You're wonderfully zealous, but please, stay away from me.

If you answered mainly (b): Congratulations—you're tactful, you're frank, and you may just save the world.

Imagine There's a Heaven

DAVID STUBBS

When it comes to the kingdom of heaven, and precisely what it entails to enter and spend eternity in this place, mainstream religion tends to be rather evasive. This is understandable. It seems to entail a state of ecstatic peace, which is a hard-to-resolve paradox. Doesn't ecstasy involve all kinds of unholy agitation? And can't peace be a bit, well, Sunday dull? Moreover, it comes with the promise of being deliriously, hedonistically, materialistically nice, which is problematic in that religious instruction and practice at their most austere require a mind-set that would actually find all that milk and honey and nectar somewhat excessive and bilious, not to say immoral and corrosive to the soul. In order to get to heaven, is it not necessary to find the very idea of the place as advertised, sticky with manna and rampant pleasure, uncongenial?

Religious zealots are not noted for their Falstaffian tendencies—although the early popes such as Benedict XII and Clement VI, whose record of sexual gorging is beyond the dreams of parody, provide an exception. One imagines, for instance, the Ayatollah Khomeini in heaven, sitting in a wooden chair against the wall, waving away all offers of goblets of nectar or concubines, looking broody and wishing he could initiate a fatwa or something, as in the fun old days on earth. Or Pope Paul VI, refusing anything stronger than a wafer biscuit and declining to join in the collective dance of joy at being in bathed in the radiant presence of the Almighty, for fear it would seem unseemly to His Holiness. Alan Bennett's Elizabeth II remarks in *A Question of Attribution*, "I suppose Heaven will be a bit of a comedown for me." So it would be for these notables, feeling like spare parts, like the vicar at the wedding disco. As for the brimstone and evangelical crowd, one suspects that for them heaven would function like an executive box at Old Trafford, enabling them to look down on hell's tormented from a comfortable vantage point. That, for them, would be the fun. Can you see hell from heaven?

In the past, well-meaning prose writers depicted the pure bliss of heaven as a place where you simply assumed a supine position and opened your mouth as a perpetual shower of sweet foodstuffs descended into your mouth. But you don't have to be a self-flagellating denier of the pleasures of the flesh to find all that a bit much—not unlike Marge in that episode of *The Simpsons* in which the family lend up at a confectionery trade fair and she, having had her understandable fill of saccharine, attempts to grab a bite from a raw celery stick, only to be instructed by a security guard, "You'll need to put some sugar on that." Is there celery in heaven?

Conversely, the biblical take on heaven is not exactly a pulse-quickening one. Let us turn to Revelations 4:

> After this I looked, and, behold, a door was opened in heaven: and the first voice which I heard was as it were of a trumpet talking with me; which said, Come up hither, and I will show thee things which must be hereafter. And immediately I was in the spirit: and, behold, a throne was set in heaven, and one sat on the throne. And he that sat was to look upon like a jasper and a

sardine stone: and there was a rainbow round about the throne, in sight like unto an emerald. And round about the throne were four and twenty seats: and upon the seats I saw four and twenty elders sitting, clothed in white raiment; and they had on their heads crowns of gold. . . . And before the throne there was a sea of glass like unto crystal: and in the midst of the throne, and round about the throne, were four beasts full of eyes before and behind. And the first beast was like a lion, and the second beast like a calf, and the third beast had a face as a man, and the fourth beast was like a flying eagle.

Leaving aside the dubious prose style of the Good Book—all those "ands," which remind one of the essay of a seven-year-old schoolboy: "Yesterday I had my birthday party and Jason came and Kyle came and Alice came and we played dead lions and we had jelly and I had a birthday cake and I blew the candles out all by myself"—does this heaven really strike anyone as any kind of fun? The everlasting company of sedentary elderly men and a random assortment of animals, none of which were your pets, and at least one of which would regard you as a tasty snack? I think I'd rather be playing dead lions. Is there alcohol in heaven?

Troublingly, heaven can be presented as an eternity spent in church, in the company of God, whose implacable ego subsists on a perpetual and universal diet of unbridled worship, as thanks for the crackerjack job He did in creating the good old world. The difference is, however, that this heaven is bathed in perpetual, Dantean blazing light, the sort that gives choirboys a headache when it pours through the west window during evensong, or not unlike the constant fluorescent light under which inmates at The Hague are incarcerated. Is there darkness in heaven?

Worship is not fun. Worship is tedious and hard work and there's precious little in it for the worshiper, as Peter Cook, playing the fallen angel in *Bedazzled*, demonstrated by perching himself atop a post box and having Dudley Moore dance around him praising him for twenty seconds or so until Moore eclared himself bored. As the realization has tacitly crept over the religious powers that be that kitsch definitions of heaven as golden, milky, cathedral-like, of a jasper and sardine hue, etc., are limited in their banality, they have shied away from such visualizations. And so we turn to movie depictions of heaven—not the myriad Woody Allen–type dream sketch scenarios involving bearded men perched on clouds with unconvincing gold-sprayed wire halos attached to their heads but, for example, Powell and Pressburger's *A Matter of Life and Death*.

This is a superb movie, in which David Niven's apparently doomed RAF pilot is spared death through a blunder in the hereafter. Eternity here is depicted as a vast affair of administration and assemblies, a celestial civil service. One new entrant is puzzled at the sight of one of the drones behind the counter. Is that his lot for eternity, he asks the chief recorder? "For some, it would be heaven to be a clerk," she replies with a very British postwar decency and high-minded complacency. This is heaven as perfect mid-twentieth-century Middle England, with everyone in their place and God absent—like Father, not to be disturbed in his study. But the drab implications of such a world are exacerbated by the monochrome in which the scenes in the hereafter are filmed. Is there color in heaven?

There's certainly color in 1998's *What Dreams May Come*, one of a hideous cycle of gruesomely sentimental Robin Williams films in which color is indeed abundant, in which the afterlife is represented by the images of his wife's artwork, which means that he is splashing about through paint—ooh, and look, there's his dog! This is heaven as New Age CGI fantasy, as wishy as it is washy—while there's only probably no God according to the bus, there's definitely no this. Thank Christ. Is there Robin Williams in heaven? If so, include me out.

Sexual pleasure was once awkwardly advanced by my old religious teacher as a simile for the bliss of heaven—think, he said, of a perfect evening you have spent with a girl. (I was in the fourth form at the time at a boys' school.) The eighteenth-century English poet Rochester in his *Love and Life* protests that the unsustainable passion of sexual love, the "live long minute," is "all that heaven allows," which implies that sex is but the teasingly dispensed sampled phials of the perpetual orgasm that is life in the hereafter. However, this is not a road of comparison church people generally like to go down. Sex and the religious life have always been problematic, as the neurotic misery passed down the centuries by St. Paul attests, as does the necessity of conceiving of Mary's virginity in order to affirm the lowly, bestial act of procreation. Will there be sex in heaven?

And so modern definitions of heaven have veered toward more abstract formulations, a state of grace beyond conventional space and time, or, according to the dictionary, "a spiritual state of everlasting communion with God . . . a place or condition of utmost happiness . . . a state of thought in which sin is absent and the harmony of divine Mind is manifest."

Here, however, is where the real problems begin. First, this is the sort of "heaven" that Christopher Hitchens compares to living in North Korea—everlasting gratitude to a Dear Leader. Then there is the problem of the dissolution of the self in heavenly bliss. Sin, disharmony, imperfection, deviation from some organic, universal, absolute perfection—this is what makes us ourselves. Our flaws, quirks, and idiosyncrasies define us. Erase them, iron them out, and we cease, in effect, to exist—we become drowned, as individuals, in some liquid other, just a part of the ecstatic ether, without the capacity for observation or reflection. Consciousness is about holding yourself in reserve and reserving the right to do so. For instance, the reason I'm reluctant to participate in happy-clappy or gospel celebrations isn't ultimately out of disbelief in the God they're serenading—even if I believed in God, I'd still want to opt out of that choral euphoria. I don't like acoustic guitars and cheery vocalizing, that's all. More of a Nick Drake man myself. Is there a jukebox in heaven?

People commonly ask why, if God exists, He allows suffering in this world. However, it is a cruel fact that in order for life to have any definition, any contrast, any light or shade, evil is necessary. So long as it is not we who are enduring its worst manifestations, it enriches our lives. So Baltimore has to be laid waste at every level in order to give us *The Wire*. Confederate soldiers have to be mown down en masse in order to give us the DVD box set of *The Civil War*. A Spanish town has to be strafed in order to give us Picasso's *Guernica*. And so forth. Is there a *Guernica* or a *Wire* or a Globe Theatre in heaven?

No; all such things, and presumably, the memory of them, will have been deleted. In their stead, is the less palatable option of the Happy Cotton-Candy Bunnies Grinned Inanely and Eternally, Their Minds Switched Off Forever, as They Forfeited All Memory of Life and Became Part of an Indivisible Organism of Collective Euphoria. Which would not sell well on DVD at all. Heaven is worse than a lobotomy. It's a wipeout. It would both suck and blow, to quote Bart Simpson, except there would be no sucking and blowing in heaven, given that there is no respiratory requirement to breathe. I guess there is no belly laughing in heaven, either.

Woody Allen once wrote that he did not want to achieve immortality through his work—he wanted to achieve it through not dying. That I can relate to. This life is where it is at. The "better place" is no such thing. Frankly, my dears, I'd rather die.

A Happy Christmas

A. C. Grayling

It does credit to our ancient forefathers that, based on their observations of the sky above them and the earth around them, they understood nature's cycles. They almost certainly had not a linear sense of time but a cyclical one, based on the return of the seasons in regular succession. Each year they knew the signs of oncoming spring, of when the hottest time of year could be expected—the "dog days" associated with the star Sirius—and the difference between the autumn of the harvest and the autumn of the fall, when all nature seems to prepare for sleep.

And they knew also what to do about the darkest part of winter, when the sun is at its furthest point south, making the days short and the nights long. Popular imagination today thinks of ancient midwinter as a time of bleakness and dearth, but this is incorrect. When almost all our ancestors were in some way involved in agriculture, winter was neither bleak nor hungry. The harvest was stored, the animals fattened or already slaughtered and salted, the larders and granaries were full, and winter was a time of rest: a time to mend tools, make clothes, sit round the fire and tell stories, sing, recite, and plan ahead. Because the days were short and darkness was the prevailing note, it was also a time of lights: to lighten the darkness, to provoke enjoyment and pleasure consistent with rich supplies of food and the leisure to enjoy them.

Somewhere among these facts, supposed to obtain in some form for thousands of years in the deep prehistory of our species, lies the origin of the historically known midwinter festivities, which, right up to recent times, constituted an extended holiday. The "twelve days of Christmas" derives from the pagan Scandinavian winter celebration, in which a great log—the Yule log—is set alight to burn for the twelve days of the holiday, celebrating the beginning of the sun's return journey from its southernmost point. The date of Christmas, though associated with earlier pagan observances of the winter solstice, is drawn from Roman festivals that were important at that time of year. In the old calendar the winter solstice fell on December 25, and in the Roman Republic right up until its end in the principate of Augustus, that date marked the end of the Saturnalia, which began on December 17 and lasted a week. In his account of the Saturnalia, Lucian wrote that during this week "everything serious is forbidden; no business is allowed. Drinking, noise, games, gambling, appointing kings and feasting of slaves, singing naked, excited clapping of hands, from time to time ducking corked faces in ice water—these are the occasions I preside over." Lucian's remarks are a clue to an earlier origin of the festival: the Greek honoring of Dionysus, also on December 25. Many Romans likewise treated that date as the feast of Bacchus, god of wine; this was the Brumalia, *bruma* meaning "shortest day."

Later, in the Roman Empire, December 25 became a universal public holiday associated with the festival of the "unconquered sun," denoting the return of the sun from its southernmost declension. This allowed several different religions in the Empire to share the same day for similar traditional reasons, their objects of worship including Mithras, the Persian deity much beloved of Roman soldiers; Sol, the god worshiped by the emperor Aurelian; and Elahgabal, a Syrian sun god. Its most important source was of course ancient Egypt, where sun worship (Ra, the sun god, was the king of the gods) and the winter and summer solstices had immense importance.

From all these Roman festivals come the traditions of exchanging gifts, decorating one's home with garlands of plants, and putting candles in trees. In the early fourth century CE the church adopted the date for its own celebration of Christ's birth, as a way of profiting from the traditional associations of the date and its surrounding days as a time of holiday.

To these elements came to be added a mishmash of others. Mistletoe and holly are Druidic sacred plants, and their use in Christmas decorations recalls the Druids' interest in solstices as especially significant occasions. Santa Claus (St. Nicholas), whose feast day is celebrated on December 6 in many European countries, and Father Christmas (a Nordic figure of legend) began as two different individuals, but they have become conflated with each other, and are now one and the same individual. Thus Santa Claus drives a flying sleigh drawn by reindeer; the notion is a pastiche.

What one learns from all this is that the darkest part of winter, which at the same time was the season of greatest plenty and leisure, was a special moment in the year for all our forebears from the most ancient times, and it is no surprise that so many religions came to observe a major feast at that period of the year. Christianity, a young and syncretistic religion drawing elements from many other faiths and superstitions that antedate it, is no different. Like its predecessors, it borrows the darkest weeks of winter for one of its own foremost festivals.

It states the obvious to remark that in the increasingly secular Western world, Christmas remains important chiefly as an opportunity for family reunions, gift giving, relaxation, and enjoyment, even for those who find no religious significance in it. Leaving aside the frantic commercialism that now dominates the season, this aspect of the tradition is potentially a good one if sincerely felt and genuinely enjoyed. Naturally enough, some find it a trial to be cooped up with uncongenial relatives, overeating and overdrinking in an overheated house, forced to watch appalling telly, and often confronted with unappetizing weather out of doors that inhibits going for a walk. Forced jollification is a nightmare, and even though the modern world is mercifully spared the traditional full twelve days, the period between Christmas Eve and New Year's Day can feel every bit as long.

For Christmas-disliking folk, the dream is a Christmas spent in a warm country where they do not celebrate Christmas. They would revel in the absence of Christmas music, decorations, and symbols, together with exhortations to spend money on trivia, ephemera, and excessive quantities of food and drink. They would be refugees from the iterated "Jingle Bells" and other carols that play on a loop in every department store, driving the staff mad. They would be escapees from the relentlessness of Christmas imagery and sentiment. They would have the sanity and fresh air of a place unconcerned to believe, or even merely pretend, that it is enjoying itself.

No such escape is open to those with young children, for whom Christmas is a bonanza of acquisitiveness and indulgence, and yet to whom we all wish to give the traditional experience of acquisitiveness and indulgence. It is in large part because of our children that Christmas has accumulated its hybrid and generally over-the-top contemporary form, together with its sentimentality and excesses. It has become a piety to approve of this, so to call it in question is to invite being called a Scrooge or worse.

For my part I think the idea of a bit of time off, which we devote to those we care about, and to whom we give tokens of our love and thoughtfulness, is a good thing. A gift that is really well chosen shows how one has thought about the intended recipient and put to work one's understanding of him or her, and one's affection. To take time to think about such a gift, and then to find and buy it or even to make it, is a real mark of love. To set aside dedicated time to be with people for whom one has deep feelings, time that is specifically for them and for the nourishing of one's relationship with them, is both a fine and a necessary act. So having a season in which we do these things is good, a genuine component of the overall good of life.

I should therefore like to see the bit of time off, which we devote to those we care about, and to whom we give tokens of our love and thoughtfulness, happen more than once a year. Perhaps 365 times a year would be a good number. But if that sounds too exhausting, not to say too expensive, then twelve times a year would do. The fact that it happens only once a year as (so to speak) an official and obligatory thing is part of the explanation why it can sometimes seem so trying, though doubtless I shall be called a Scrooge for saying so. But it is also the reason why, when I wish those I meet "a happy Christmas," I really mean it.

Beloved Buzzkill

ALLISON KILKENNY

What's the harm, I thought, if good people believed in a fictitious male figure who lived somewhere above here and listened to their hopes and prayers? After all, Santa seems harmless enough—so why not God? It wasn't like my believer friends and family were waging holy wars, bashing gay people, or burning crosses on black people's lawns. Generally speaking, they were smart liberals who also happened to believe Jesus turned water into wine, which to me is weird.

Maybe people think I'm weird, I thought. I was a vegan atheist growing up in the Midwest. I was practically an alien.

So I stayed quiet about the non-believer thing.

My silence lasted through college. It's remarkably easy to avoid existential confrontations and lengthy theological discussions when one's schedule primarily consists of getting shitfaced in between panicked, last-minute cram sessions. I graduated and moved to New York City, where I started working at a bookstore. Though I was surrounded by thousands of books harboring millions of ideas, God still didn't come up in conversation. Co-workers and customers like to keep chatter light and pleasant during working hours: the crappy weather, plans to drink next weekend and mate with the new register girl, and so on.

In fact, God didn't come up until I unexpectedly fell in love.

It happened on the day I wore my Superman T-shirt. There I was, probably staring off into space, which is how I spent 90 percent of my day at work, when a young man's face filled my vision. He was saying something. I could tell because his lips were moving.

"I have your shirt." That's what he said.

"What?" I replied.

"Well, not that one. Mine's a guy's shirt—it's the guy's *version* of that shirt. Nothing. I—I'm sorry," he stammered.

Then he was gone.

I would later learn the young man's name is Jamie, and he's not always that inarticulate. In fact, he's an extremely smart comedian who loves Bill Hicks (swoon), politics, and considers himself a bit anti-authoritarian.

We turned out to be terrible booksellers, although we were excellent at spending every day huddled at the main information desk, talking about absolutely everything. We expelled chatter in the excited, breathless way children do at sleepovers when the lights are out, when it feels like there's just not enough time in the universe to share all the ideas buzzing around one's head. And when we actually started dating, our productivity level plummeted further. Our managers were furious. They did everything in their power to separate us, but like a retail Romeo and Juliet, we defied the fascist restraints of our overlords and rendezvoused behind book stacks to chat for hours.

And God never once came up. Until, that is, Jamie and I were attempting to coordinate our holiday break schedules so I could visit his hometown to celebrate Christmas with his family. He was trying to explain to me the distinctly modern faith amalgam of his familial unit in order to brace me for the eclectic hybrid of Chanukah-Christmas celebrations.

"Well, my dad's Jewish, but really more culturally Jewish. He's not *Jewy* Jewish. Someone usually reads from the Torah, but only for a couple minutes. Then my mom's Christian, but we never went to church much. She prays sometimes, but not in the crazy speaking-in-tongues way," he said, all the while glancing at me, seemingly waiting for me to chime in with my own set of beliefs.

When I said nothing, he continued, "Personally, I don't consider myself Jewish or Christian." I perked up a bit here. Could Jamie be an ath—"But I do believe there's *something*."

My heart sank. "Something?"

He started gesturing with his hands, the palms up. "Yeah, not like a judgmental psychopath God, but *something*. I don't know. How else can you explain all the beautiful shit in the world?"

Biology. Evolution. Physics. A thousand explanations immediately flooded my brain, but I just smiled and nodded. After all, I liked Jamie a lot, and I didn't want to come across as a shrill heathen. I could pretend to believe a little longer.

But Jamie saw through me. "You're not . . . like . . . agnostic, are you?"

I felt nauseated. "Um, I'm . . . an atheist."

I thought Jamie was going to faint. He turned pale. His eyes widened. I can't be sure, but I think he twitched.

One of us changed the subject.

When we put in for our vacations, we ended up getting more time off than initially expected. Jamie planned a pre-holiday celebration detour to Niagara Falls, which would take us hours out of the way. Nothing could have been better.

It's amazing how love puts a shiny coat on the worst circumstances. Traffic becomes a conversation-elongater. The crowds of Niagara tourists are the Multitudes Who Shall Witness Our Love. The freezing weather makes huddling together for warmth a delightful necessity.

Everything was amazing as we shivered side by side, gazing down at nature's majesty. Truly, it was one of those moments where words seem utterly inadequate. I felt small and humble, and all I could do was watch the roaring, voluminous falls crash into the huge boulders at the base, kicking up clouds of mist.

"See, like this," Jamie said. "How can you explain something beautiful like this? God must have done this."

I almost nodded again, if only to quickly squelch the conversation and return to the serenity of silence. But then something occurred to me. By believing that "God did it," Jamie was missing out on a much more interesting story, and I was guilty for facilitating his ignorance.

No one could blame a person like Jamie for simply being somewhat ignorant about certain things. However, a person like me—someone *who knows better*—is the worst kind of ignorance-enabler. How could I claim to love this person if I insisted on allowing him to believe something that simply isn't true?

"Actually . . . ," I gently began.

And so I told Jamie about how Niagara Falls was created when the Wisconsin glacier receded during the last ice age.

To my great surprise, Jamie didn't immediately scream, "*Burn her!*" and tie me to a stake. At first he seemed thoughtful, and then the floodgates opened.

He wanted to know everything. How did flower petals get their color? How did the planets form? What is a star? I did my best to answer his questions to the fullest extent of my ability. Sometimes I didn't know what to say, so I pointed him toward the experts: Richard Dawkins, Neil deGrasse Tyson, and on and on. When the experts could take the answer only so far, I explained to Jamie that the limits of human understanding don't prove God stands on the horizon.

Before there was an understanding of gravity, humans believed God explained why an apple falls down but never up. When Newton provided a satisfactory answer, God took a step backward. Each time scientists made another scientific discovery, God took another step backward.

Whenever something occurs that cannot be understood by science in that moment, the natural inclination is to throw up our hands and surrender: "God did it!"

A. C. Grayling made the observation that humans used to believe in multiple gods that lived in the trees, water, and bushes. They made the wind and controlled the tides. When human explorers searched all the trees and swam in the water and found no gods, their mythology changed.

Suddenly the gods didn't live in the trees but resided in the mountains. And when human explorers walked high into the mountains and found no gods, the mythology relocated again. The gods lived in the sky, and now that humans have explored space and found no God, the Creator has taken on slightly more metaphysical properties. Perhaps God is the universe, or some laissez-faire force.

Of course, these are all desperate attempts at explaining why humans are not alone. The rationality is fear-based, which is understandable, but at this point—with all we know about science and the universe—inexcusable.

I told Jamie all of this, and at times I did feel a little like some monster adult telling a child there's no Santa Claus. Jamie was (and is) a good person, and he used religion in the best possible ways: to help people and to act compassionately. Would I reverse that behavior by telling him these things?

On our way from the falls to his family's house, I asked him how he felt about viewing the world scientifically rather than religiously. Jamie didn't speak for a long time. He looked thoughtful, and for a moment I was terrified. Did he blame me for ruining the fairy tale?

Did he reject my explanations? Had I become the buzzkill I had always fought so hard not to be?

"In a way," Jamie began, "I think the world is more beautiful this way. I mean, before, I didn't know the story behind how things work, but there's so much . . . *more* to it now. And without believing in a heaven, or a God, and that stuff, it makes life so much more *valuable* now."

"Valuable?" I asked.

"Yeah. I mean, we have to take care of each other because God's not going to do it for us," he said.

To my great relief, telling Jamie about my atheism and helping him to recognize his own atheism didn't result in a devaluing of life. Rather, it enhanced his view of the world. Instead of believing some divine matchmaker had thrust us together against our will, Jamie now believes we had the extraordinary good fortune of finding each other in a sea of 6 billion people. And that's just *so* much more interesting, and amazing, and something for which we should forever be grateful.

The best possible decision I ever made was to "come out" as an atheist. Now I know that it's not inherently rude to combat the ignorance that is inherent in religion. In fact, I now know education is—by default—the enemy of religion, which loathes dissent and curiosity. My silence was enabling the worst, most corrosive aspects of religion.

Jamie is now one of the most outspoken atheists I know. I've witnessed him educate believers and usher them into a world of rationality. Sometimes I think about how many recovering God-believers we've helped, and how many would still subscribe to antiquated theology had I kept my mouth shut and never shared what I know with Jamie.

One of the happy by-products of love is the ability to grow together. By embracing reason, Jamie and I did just that. To us, this became our new spirit of Christmas. It was no longer a holiday meant to worship the birth of a now dead Palestinian who once claimed to be the Son of God but who is now used as a marketing ploy by huge corporations to sell more Xbox consoles. Christmas became a time to celebrate human connection, whether that is with family or friends, and to remember that sometimes the gift of knowledge is the best present you can give.

Stay Away from My Goddamn Presents

JAMIE KILSTEIN

Christmas is my favorite holiday ever. There. I said it. Judge me all you want, but it's out there in the ether. No taking it back. I am an anti-capitalist atheist, and the holiday that represents everything I loathe about our greedy, consumer-based, fairy-tale-worshiping country is my favorite holiday—hands down.

Really, Halloween? Candy is the best you got? Call me when you have a giant tree of awesomeness and lights illuminating *stacks of presents*! Disgusting.

I should have known any holiday that involves shameless bribery would turn out to be a scam. Kidnappers never trick anyone into their van with promises of peer-reviewed studies on Darwinian evolution. Worst. Kidnapping. Ever. But Christmas—with its awesome songs—could have dragged me into its van anytime.

That's why, when I fell in love with a girl who tried to take Christmas from me, I had to put up a fight and remind myself it is unbecoming to strike women. Historically, people have made lots of mistakes for pretty girls—they've died, killed, and destroyed bands. No way this one was going to take away my presents and whatever the hell else

Christmas is supposed to mean. (Something about a birthday and not letting gays marry? I don't know. I never went to church.)

Before I met the vixen who wanted to rob me of my childhood, I believed in God for all the right reasons: when I wanted shit, I could pray to him; when I felt bad about doing something shitty, I would assume he would fix it; and, um . . . I don't know what else . . . *presents*!

God was my fall guy, my wingman, my wish granter, you name it. Sure, if people really think He is almighty, we should probably be praying for Him to fix up that whole Middle East boom-boom thing, but if God is almighty, He can probably do that while simultaneously helping me find my keys, right? I need my keys, people.

It was a pretty great relationship, me and God. If I saw a homeless person and felt bad, I knew God would never let someone starve, so I didn't need to do anything. (Africa who?) If I wanted to start smoking again, I knew God had a plan. Maybe if I started smoking again it was because I was meant to go into the corner store on Forty-second Street to buy a pack of smokes, and at the same time it was being robbed, and I would be there to stop the robbery! Who else would stop it? Nobody! Just me—a tiny artist who is afraid of confrontation.

I didn't need to go to church because I wasn't a conformist, man. I was spiritual but not religious. Which I think just means I'm pro-choice, church bores me, but I don't want to go to hell, and was keeping my options open. It was sweet living.

But then Allison came into my life with her *facts*. And *words*. That she read in *books*. The nerve of some people.

I fell in love with Allison pretty quickly. We were working at a bookstore, which is where smart, pretentious people who can't get real jobs work. It's retail, but we can read, so we assumed we were better than everyone else.

She listened to cooler music than I did, knew who Noam Chomsky was, liked Bill Hicks, and wanted Dane Cook dead. She was the smartest girl I had ever met.

We pretended we were just going to be friends. The façade lasted about a day. I remember during our time as "friends" we would meet up before the morning shift at the bookstore because "friends" love meeting at Central Park, hung over, at 6:00 a.m. before an eight-hour workday. I don't care what joggers tell you—nobody wants to be up at six in the morning.

When we weren't meeting in the park, we talked nonstop at work, huddled behind the main information desk. During one of those meetings, we did what most couples do: ask each other scary political questions and hold our breath, hoping the person we were just making out with doesn't suddenly drop some horrible bombshell of crazy, like "Yeah, I know gays are *technically* people . . ." *Oh God, please don't ruin this.* I believe it was our third date, which is when questions like this come into play. Not during the first date. If on the third date the person turns out to be crazy, at least you got some make-out time in. But the third date is all romance and talk about your thoughts on abortion and the occupation of Gaza.

We matched up pretty nicely. We were both Democrats who hate the Democratic Party, both vegans and animal rights activists. Both of us despised the war, assumed Dick Cheney was a villain out of a cartoon, and cried at the movie *Serendipity,* with John Cusack . . . Okay, that was just me, but only because Allison doesn't have a soul.

The conversation then turned to religion. I was pretty confident everything would be okay. Allison and I are pro-gay-rights and pro-women's-rights, know about evolution, and think that maybe bombing Palestinian children isn't worth Jesus coming back on a horse, or whatever weird Armageddon shit they had planned. We agreed on all of that! All was right in the world! But then this happened:

Jamie: So you're agnostic?
Allison: No.

Jamie: Wait . . . [*nervous laugh*] . . . you're not an atheist, are you?

Allison: Yeah. I'm an atheist.

I wish I could write this without having to tell you what my actual thoughts were, but for the sake of honesty, here we go. Allison told me she was an atheist, and my first thought—as an adult, as a progressive, non-religious adult—was, word for word, this:

Jamie's brain: You can't be an atheist . . . you'll burn in hell.

That was not fun to write. But it's what I thought. I had never been to church. Unlike Allison, I never went to Sunday school, was never threatened with burning in hell, never had to use reason and courage to overcome these fears, but I grew up in America. And in America, atheists are the most hated minority. Really. Look it up. And after you look it up, start spreading rumors about the Quakers, because this really is not fair. We are somehow the most charitable and most hated group at the same time.

I had never heard the word *atheist* used by someone I knew. It just had such awful connotations to me that when she said the word, my heart stopped.

Even though I never really thought about hell, it still scared the shit out of me. (Funny how people believe in heaven because that part sounds really fun, but when it comes to hell and suffering, most of us are like, "Oh, that's probably not true.") I remember thinking: *Even if you are an atheist, just shut up and say you're agnostic. Trick God with me. What does that guy know, anyway? He created Sarah Palin and Michael Bay. We all have our off days.*

I tried to talk her out of it, and in doing so—for the first time in my life—I became devoutly religious. This was somebody I cared about, and yes, all evidence and reason say that there is no God, but I loved this girl and could not take the chance. Sorry, evidence and reason!

Evidence and Reason: But Jamie, chances are there is no giant, red, muscular dude in a lake of fire, who chases people with his pitchfork. What are you, nine?

Jamie: Shut up, Evidence. *You're* nine!

Evidence and Reason: C'mon man, that doesn't even make sense. We don't have an age.

Jamie: You're nine *and* stupid.

Evidence and Reason: We'll be back once you've calmed down.

Previously, I never understood what "saving" people meant, but fearing for Allison's eternal soul made me understand how some evangelicals can really believe they are saving people from burning in hell for eternity. That explains their moxie.

I was turning into the people I mocked. I kept trying to convince Allison to say she believed in God. She didn't even have to really believe it—just say it. I told her I didn't want to be in heaven alone (I am twenty-seven years old, and those words came out of my mouth). Anytime she pointed out the crazy mistakes in the Bible, I told her God probably had nothing to do with the Bible. God laughs at people who read the Bible, but he loves us. That's how she and I had found each other.

Jamie: Just shut up and stop being so stubborn, and pretend to believe in something, so you don't have to burn forever.

Allison: No.

I couldn't understand why she didn't see the romance in this. If you lined up all the facts, Allison and I should have never met. Clearly, it was God who'd put us together. She was being so ungrateful!

It happened like this.

The day before Allison started working at the bookstore, I was going to quit. I went into my manager's office and bravely told him that I was leaving. Then he told me I couldn't leave, and within seconds I caved because I am weak.

The next day Allison started working at the bookstore! Before Allison got the bookstore job, she was going to join the Peace Corps, but changed her mind at the last minute. Slam dunk. God did it.

I would only begin to understand much later that my thinking during this time was completely selfish. I was thinking about it like God was only looking out for Allison and me. In reality, if God were playing matchmaker, that means he just took a very qualified girl away from the Peace Corps . . . to work retail. Allison could have saved Somali children, but no, no, God wanted me to screw the new girl. That doesn't seem right. I get laid, but those kids are pirates now.

But belief like that makes you feel special. It's like God is watching you. We like to pretend we are our own mini *Truman Show*, forgetting that Truman wanted out. When people say, "It's in God's hands," that just means they don't want responsibility for what's about to happen.

Allison refused to relent. I stopped trying to convert her. When the holidays rolled around, I took Allison to Niagara Falls because I needed to distract her from the fact that I was broke and couldn't afford presents (thanks, retail). She already summed up that story better than I can, but I'll add this: I stopped believing in God that day, but I started to believe in life and love.

I liked Christmas for all of the wrong reasons. Gifts and shiny lights will eventually get old or burn out, but love is something that stays with us until our last breath. I don't need a man with a sled and beard to tell me that. I don't need to think that God put Allison and me together like some perverted voyeur. There are more than 6 billion people on this planet, and if you find the one person you want to spend the rest of your life with—if you find someone you want to spend more than twenty minutes with—it's a miracle. You did that—no one else. That, to me, is romance.

The First Honest Christmas Round-Robin Letter

Julian Baggini

Dear All,

It's been a year of transformation and rejuvenation for the Goodhead family! Everyone—well, nearly everyone—is so looking forward to a Christmas that promises to put the icing on the top of our cake of annual achievement!

It all started last Christmas, to be precise, at around three-thirty on December 25, when I brought out the flame-licked pudding.

"Who's for a bit of Crimbo pud?" I asked, not expecting the answer I got from our eldest, Toby.

"I don't think I can take any more of this," he sighed, draining his large glass of brandy. The room fell silent. "I mean, none of us even likes Christmas pudding, and even if we did, we're all too stuffed for it anyway."

For a short moment we all took on the appearance of startled squirrels. We had all thought his was an existential shriek, a declaration that our imperial Santa had no clothes. But perhaps all he could take no more of was lunch. We felt the embarrassment of people who had just agreed that S&M was indeed great, only to realize that the topic of conversation was in fact M&S.

Sensing our uncertainty, Toby spelled out what he had meant. "Think about it. We pretend we're a family of rational freethinkers, with no time for supernatural nonsense and pointless social conventions. Yet look at us. Once again, we've wasted money on pointless presents nobody really wants. I mean, Dad, surely you didn't think I'd read *Is It Just Me, or Is Every Book a Rip-Off of Something Else?* Why didn't you give it straight to Oxfam and cut out the middleman?"

Gerald was indignant. "Talk about the kettle calling the pot black!" he said, a mix-up so typical no one bothered to correct him. "What about the book you bought me?"

"Excuse me, but just because it's piled up next to all the unfunny, rip-off stocking fillers, that doesn't mean *The Atheist's Guide to Christmas* is one of them. Anyway, it's not just the crap presents. I can't believe how much rubbish you've bought from the supermarket. Under what conceivable circumstances would any half-civilized human with a palate that has not been destroyed by substance abuse choose to bake frozen mushroom vol-au-vents or open a jar of apricots in brandy syrup?"

"They are organic," I said, as though it made a difference.

"What annoys me the most," continued Toby, warming to his theme, "is that once again you've bought a jar of piccalilli. I don't even know what piccalilli is. Do you?"

The room was silent.

"I agree with Toby," said Clara. "But let's face it, the excess is nothing compared to the hypocrisy. Why do we bother pretending that we're one big harmonious and happy family, spending a day of love and laughter together? You two," she said, wagging her finger dismissively at me and Gerald, "pretend for the sake of us three, as though we were still kids, and we pretend for the sake of you, as though you were old and senile."

"For Christ's sake," interjected Joshua. "It's bad enough you can't stand Dad—you don't have to rub his nose in it."

"You can't stand me?" asked Gerald, sounding more baffled than hurt.

"It's not you," said Clara. "It's both of you. And don't look so pathetically crushed. It's you who always told us the importance of speaking the truth when we were growing up. If you don't like the cake, don't give out the recipe."

"If you're so keen on a familial truth and reconciliation council," said Joshua, "how exactly do you intend to put it into practice?"

That, in short, is the question we have been trying to answer ever since. And twelve months later, I think we have reached our verdict. The simple principle we have tried to follow is to do nothing that defies or obscures the truth. As rationalist atheists, we have always had this as our official policy. But Toby's brave candor made us realize that we had not in fact been living by it.

Obviously, this means that, as usual, we will be singing no carols and going to no church services. But this year we're going much further. Toby and Clara have honestly admitted that they don't actually like us very much and that they'd rather please themselves this year. We applauded their commitment to truth and honesty and gave them our blessing. So Clara will spend Baby Jesus Day (as Gerald always calls it) by herself in her flat in Bermondsey, as though it were a day like any other. She told us she thought it would be liberating, like much of the crying she has been doing recently. We don't know what Toby will do, as he decided that since he wasn't pretending to like us anymore, he felt no obligation to keep us up to date with his movements.

Joshua offered to join us, but we turned him down, as we didn't want him to do it just because he felt sorry for us. Charity is for strangers, not your own parents. So he's going to volunteer at a soup kitchen. "Maybe the people there will appreciate what Christmas is really about," he said. Sometimes his humor is so deadpan!

If I'm honest, I expect it will feel a little odd not to have any presents to open or give on Christmas morning, and not to see the kids. But we won't be complete misery-guts! We've ordered some organic free-range duck breasts (hardly worth buying a whole bird for just the two of us) and will open a decent bottle of wine. And it will be nice not to have to spend the whole day cooking and washing up. It's not often Gerald and I get to spend a free day together with just ourselves to please, and I'm sure we'll remember how to do it when the time comes.

Everyone seems happy with this arrangement, except Joshua. He seems excessively attached to the irrationalities of culture and tradition. "The rituals of the calendar provide some variations of color to the passage of time," he says—rather pretentiously, we think, as though that made it all right. He says that social rituals work precisely because we do not just choose to do what we think is best, but go along with whatever is usually done. They lose their meaning when we adapt them to suit our own beliefs about what is right and rational. "That's the kind of logic that leads to Latin mass and belief in transubstantiation," Gerald told him. When Joshua protested that it had led him to no such thing, Gerald simply said, "Well, I don't expect you to be consistent in your irrationality."

So that's our big news. Life in the Goodhead family is more honest, open, and truthful, and that has to be a good thing. That means we have also rationalized our round-robin mailing list, of course, removing all the people we never see and don't miss, as well as those we don't much like. So, Tom and Barbara, if you are passing by over the holiday period, do drop by. Please.

There Are No Atheists in the White House

Paul Krassner

It was God who instructed Bill O'Reilly to consider every utterance of "happy holidays" to be a verbalization of "the war on Christmas." Whenever anybody claims that God talks directly to them, I think they're totally delusional. George Bush is no exception. Not only was he told by his senior advisor, Karen Hughes, not to refer to terrorists as "folks," but Bush was also being prompted by God Him-Her-or-Itself: "God would tell me, 'George, go and end the tyranny in Iraq.' And I did." As if he were merely following divine orders.

In July 2003, during a meeting with Palestinian prime minister Mahmoud Abbas, Bush told the newly elected leader, "God told me to strike at Al-Qaeda and I struck them, and then He instructed me to strike at Saddam, which I did. And now I am determined to solve the problem in the Middle East. If you help me, I will act, and if not, the elections will come and I will have to focus on them."

Abu Bakar Bashir, an Islamic cleric and accused terrorist leader, has said that "America's aim in attacking Iraq is to attack Islam, so it is justified for Muslims to target America to defend themselves." That's exactly interchangeable with this description of Bush by an unidentified family member, quoted in the *Los Angeles Times*: "George sees [the war on terror] as a religious war. His view is that they are trying to kill the Christians. And the Christians will strike back with more force and more ferocity than they will ever know."

Indeed, General William Boykin, deputy undersecretary of defense for intelligence, said that "George Bush was not elected by a majority of the voters in the United States, he was appointed by God." Discussing the battle against a Muslim warlord in Somalia, Boykin explained, "I knew my God was bigger than his. I knew that my God was a real God and his was an idol."

Apparently, religious bigotry runs in the family. Bush's father, the former president: "I don't know that atheists should be considered citizens, nor should they be considered patriots. This is one nation under God." And before him, there was Ronald Reagan: "For the first time ever, everything is in place for the Battle of Armageddon and the Second Coming of Christ." Not to mention Reagan's secretary of the interior, James Watt, responsible for national policy on the environment: "We don't have to protect the environment—the Second Coming is at hand."

In 1966, Lyndon Johnson told the Austrian ambassador that the deity "comes and speaks to me about two o'clock in the morning when I have to give the word to the boys, and I get the word from God whether to bomb or not." So maybe there's some kind of theological tradition going on in the White House.

But if these leaders are *not* delusional, then they're deceptive. And in order to deceive others, one must first deceive oneself until self-deception morphs into virtual reality. In any case, we have *our* religious fanatics, and they have *theirs*. In September 2007, on the eve of the sixth anniversary of 9/11, Osama bin Laden warned the American people that they should reject their capitalist way of life and embrace Islam to end the Iraq war, or else his followers would "escalate the killing and fighting against you."

Bush once proclaimed, "God is not neutral," which is the antithesis of my own spiritual path, my own peculiar relationship with the universe, based on the notion that God is *totally* neutral—though I've learned that whatever people believe in works for them.

My own belief in a deity disappeared when I was thirteen. I was working early mornings in a candy store in our apartment building. My job was to insert different sections of the newspaper into the main section. On the day after the United States dropped the first atomic bomb on Hiroshima, I read that headline over and over and over again while I was working. That afternoon, I told God I couldn't believe in him anymore because—even though he was supposed to be a loving and all-powerful being—he had allowed such devastation to happen. And then I heard the voice of God:

"Allowed? Why do you think I gave humans free will?"

"Okay, well, I'm exercising my free will to believe that you don't exist."

"All right, pal, it's your loss!"

At least we would remain on speaking terms. But I knew it was a game. I enjoyed the paradox of developing a dialogue with a being whose reality now ranked with that of Santa Claus. Our previous relationship had instilled in me a touchstone of objectivity that could still serve to help keep me honest. I realized, though, that whenever I prayed, I was only talking to myself.

The only thing I can remember from my entire college education is a definition of philosophy as "the rationalization of life." For my term paper, I decided to write a dialogue between Plato and an atheist. On a whim, I looked up atheism in the Manhattan phone book, and there it was: "Atheism, American Association for the Advancement of." I went to their office for background material.

The AAAA sponsored the Ism Forum, where anybody could speak about any ism of their choice. I invited a few acquaintances to meet me there. The event was held in a dingy hotel ballroom. There was a small platform with a podium at one end of the room and heavy wooden folding chairs lined around the perimeter. My favorite speaker declared the Eleventh Commandment: "Thou shalt not take thyself too goddamned seriously." Taking that as my unspoken theme, I got up and parodied the previous speakers. The folks there were mostly middle-aged and elderly. They seemed to relish the notion of fresh young blood in their movement.

However, my companions weren't interested in staying. If I had left with them that evening in 1953, the rest of my life could have taken a totally different path. Instead, I went along with a group to a nearby cafeteria, where I learned about the New York Rationalist Society. A whole new world of disbelief was opening up to me. That Saturday night I went to their meeting. The emcee was a former circus performer who entertained his fellow rationalists by putting four golf balls into his mouth. He also recommended an anti-censorship paper, the *Independent.*

The next week, I went to their office to subscribe and get back issues. I ended up with a part-time job, stuffing envelopes for a dollar an hour. My apprenticeship had begun. The editor, Lyle Stuart, was the most dynamic individual I'd ever met. His integrity was such that if he possessed information that he had a vested interest in keeping quiet—say, corruption involving a corporation in which he owned stock—it would become top priority for him to publish. Lyle became my media mentor, my unrelenting guru, and my closest friend. He was responsible for the launch of my irreverent magazine, the *Realist.* The masthead announced, "Freethought Criticism and Satire."

In the words of the late Jerry Falwell—who once said that God is pro-war—"If you're not a born-again Christian, you're a failure as a human being." We salute, then, a few *successful* human beings:

- The individual who placed the winning bid of $1,800 on eBay for a slab of concrete with a smudge of driveway sealant resembling the face of Jesus.
- The man who tried to crucify himself after seeing "pictures of God on the computer." He took two pieces of wood, nailed them together in the form of a cross, and placed it on his living-room floor. He proceeded to hammer one of his hands to the crucifix, using a fourteen-penny nail. According to a county sheriff spokesperson, "When he realized that he was unable to nail his other hand to the board, he called 911." It was unclear whether he was seeking assistance for his injury or help in nailing his other hand down.
- The Sunday school teacher who advised one of his students to write on his penis, "What would Jesus do?" Presumably, "masturbate" was not considered to be the correct answer.
- And, of course, the anonymous authors of the following quotes from various state constitutions. *Arkansas:* "No person who denies the being of a God shall hold any office." *Mississippi:* "No person who denies the existence of a Supreme Being shall hold any office in this state." *North Carolina:* "The following persons shall be disqualified for office: First, any person who shall deny the being of Almighty God." *South Carolina:* "No person shall be eligible to the office of Governor who denies the existence of the Supreme Being." *Tennessee:* "No person who denies the being of God, or a future state of rewards and punishments, shall hold any office in the civil department of this state." *Texas:* "Nor shall any one be excluded from holding office on account of his religious sentiments, provided he acknowledge the existence of a Supreme Being."

Rick Warren, pastor of America's fourth-largest church, told his congregation, "I could not vote for an atheist because an atheist says, 'I don't need God.'"

In 2006, the Secular Coalition of America offered a $1,000 prize to anyone who identified the highest-ranking non-theist public official in the country. Almost sixty members of Congress were nominated, out of which twenty-two confided that they didn't believe in a Supreme Being but wanted their disbelief kept secret. Only Pete Stark admitted that he was a non-believer, and in 2007 he became the first member of Congress ever to identify himself publicly as a non-believer.

In the week following that announcement, he received over 5,000 e-mails from around the globe, almost all congratulating him for his courage. "Like our nation's founders," he stated, "I strongly support the separation of church and state. I look forward to working with the Secular Coalition to stop the promotion of narrow religious beliefs in science, marriage contracts, the military, and the provision of social services." In 2008, he was elected to his nineteenth term with 76.5 percent of the votes.

In the 2008 primaries, three presidential wannabes raised their hands during a Republican "debate" to signify that they didn't believe in evolution, although one of them, Mike Huckabee, admitted, "I don't know if the world was created in six days, I wasn't there." He has also said, "If there was ever an occasion for someone to have argued against the death penalty, I think Jesus could have done so on the cross and said, 'This is an unjust punishment and I deserve clemency.'"

It was a pleasant surprise when Barack Obama acknowledged "non-believers" in his inauguration speech. However, I don't exempt my fellow atheists from criticism. I view as foolish those believers and skeptics alike who are waging a battle against the teaching of meditation in publicly funded schools, as though slow, deep breathing is necessarily and automatically a religious practice. What's next, forbidding the teaching of empathy because that's what Christians and Jews are supposed to practice?

Similarly, I ridicule China's atheist leaders for banning Tibet's living Buddhas from reincarnation without permission. According to the order, issued by the State Administration for Religious Affairs, "The so-called reincarnated living Buddha without government approval is illegal and invalid." The regulation is aimed at limiting the influence of the Dalai Lama, even though China officially *denies* the possibility of reincarnation. (I used to believe in reincarnation, but that was in a previous lifetime.)

China is a Big Brother, slave-labor-driven, human-rights-violating, Maoist dictatorship, from which the United States borrows trillions, then proceeds to purchase "Made in China" American flags, poisoned food, and lead-painted Christmas toys. America remains a living paradox, where our citizens are force-fed deceit and misinformation so that we can continue to fund inhumane and illegal activities—even though the revolution was fought because of taxation without representation—yet I live in a country where at least I still have complete freedom to openly condemn the government, the corporations, and organized religions that continue enabling each other to reek with corruption and inhumanity. I'm truly grateful.

"Thank you, God."

"*Shut up, you superstitious fool!*"

ARTS

When a comparatively simple, straightforward mathematical expression turns out to correspond to aspects of the natural world . . . believe me, almost every scientist who's experienced that feels the same kind of reverence and awe that we bring to great art and great music.

—Carl Sagan

An Atheist at the Movies

David Baddiel
and Arvind Ethan David

Cinema covers the waterfront of humanity. That's what you would think. Over the course of its century or so of life, cinema seems to have given us films covering every genre and subject matter imaginable, and had every possible type of hero. Slum-dwelling millionaires and Australian Scots warriors both get to display personal courage in the service of love; Chinese adulterers and gay cowboys are united in the battle against their own desires. Geriatrics are as free to get naked, fly into space, and fall in love as are teenagers—who also get to re-create Austen and Shakespeare and have underage sex, often at the same time. Pigs can learn how to be dogs; dogs come home; dinosaurs come to life and seek meaning in Los Angeles; and aliens train for the Tour de France while blowing up the White House.

In all this glorious diversity, however, there's one theme, and one class of protagonist, that seems to be strikingly absent from the canon. Given the title of this book, and indeed this chapter, we think you can probably guess what it is. You try: name three movies with explicitly atheist protagonists or themes. Not documentaries or TV programs, but proper, popcorn-accompanied, multiplex-playing movies. Take a minute, put the book down, and have a go.

No, seriously: put the book down. Think about it.

Now—how many did you come up with? You see our problem. The problem is they don't exist. Seriously: there are barely any dramatic feature movies that have an explicitly atheistic hero or theme.

So few, in fact, that even the Atheist Film Festival (http://atheist filmfestival.org), after running an Internet-wide competition for suggestions, has only been able to settle on three titles: *The Root of All Evil* and *Deliver Us from All Evil* (both of which are documentaries, a genre in which there is an honorable lineage of atheist work, most recently joined by Bill Maher's *Religulous*) and *The Life of Brian*, which, while a fantastic flick and a brilliant satire of institutionalized religion, is not even slightly atheistic—it's just a very naughty boy.

Looking further afield, one can find a few examples of movies that are at least based on work that is explicitly atheistic in its concerns. The 1992 sci-fi film *Contact*, based on a novel by America's late "atheist in chief," the scientist Carl Sagan, certainly has an atheistic protagonist in the feisty astronomer Ellie (Jodie Foster), determined to prove that the only thing in the heavens are little green men.

The movie proceeds, however, to sell out her character, the book, and its author (who died of cancer before the film was completed, and who remained resolutely and courageously atheistic even on his deathbed) by giving Ellie a pseudo-religious moment of conversion at its climax. What in the book is a complex, but still shocking, mathematically rooted revelation of our holistic interconnectedness is in the film reduced to a hammy moment where a kindly father-figure alien appears from the sky sprouting platitudes. This allows the story's religious spokesman, the evangelically horny Palmer Joss (Matthew McConaughey), to proclaim victory over his erstwhile atheist bedmate, and middle American cinema-goers a victory over scientific reason.

An even more blatant neutering is given to the film adaptation of Philip Pullman's novel *The Golden Compass*—the first of his brilliant and subversive young adult trilogy, *His Dark Materials.* The books were controversial bestsellers because of their explicitly anti-God and anti-religious subtext, taking *Paradise Lost* and retelling it to young people through the lens of fantasy-adventure. The vapid film adaptation ripped out the subtext, leaving a story in which events unfolded for no reason and without import. The kids weren't fooled and stayed away in droves, causing the film to fail and ultimately bring down the studio behind it. Truly, we atheists are a vengeful god.

Other titles that crop up in lists of atheistic films stand up little better when subjected to analysis. Bergman's *The Seventh Seal* pops up a lot on atheistic film fave lists, but while that movie definitely plays with ideas of faith and metaphysics, it does so within an explicitly theistic framework.

Horror movies are often thrown up as potential candidates, but while many of them—*The Exorcist*, *The Omen*, and *The Wicker Man*, to take the classics—are on some level anti-religious, they also by definition explicitly accept God's existence, since you can't believe in the Devil if you don't accept his Father. These films use the idea that earth is the battleground 'twixt heaven and hell as an excuse for bloody mutilation and creative carnage, with all the profitable titillation that brings in for the theaters. This strategy also explains, even more so, the success of *The Passion of the Christ.*

What are we left with, then? What about those perennial Christmas movies of doubters and unbelievers? Surely Scrooge in *A Christmas Carol* is an atheist hero; "Bah, humbug!" is hardly a statement of belief. Or how about James Stewart in *It's a Wonderful Life*, wishing to end it all, hoping for the peace of oblivion, not everlasting salvation? In both films, though, supernatural god-agencies, angels and ghosts, intervene to show the doubters the error of their ways, and the films end in an explicitly God-fearing world. Bah humbug indeed.

About the only mainstream film that we could think of with an integral atheist hero and message at its core is the dramatization of the Scopes monkey trial, *Inherit the Wind*. Brilliant though this film is, and with its hero's explicitly atheistic point of view, it is more akin to a documentary—being a fairly careful re-creation of the events of that landmark trial—than a true fictitious piece of cinematic myth.

Also, it's like fifty years old and no one alive has heard of it, still less seen it.

At this point in our survey, the two of us, filmmakers and storytellers of atheistic persuasion, were beginning to get a little concerned. One of the key functions, perhaps the prime function, of cinema is to create myths of meaning; to allow people of all castes and creeds, backgrounds and persuasions, to see themselves as they might be, to feel not alone, to see metaphors of their own life and ways of being, to have their worldviews challenged and expanded.

The great progression in mainstream filmmaking over the past twenty years is that it has become more diverse—that it now fulfills this function for a greater multiplicity of audiences. The glorious diversity we describe in our opening paragraph mainly describes films made since the late nineties, which is when Hollywood, anticipating the shifting economic centers of the world, started to cater to emergent audiences. Now more people than ever before get to see themselves writ large on the silver screen, get to think of themselves as heroes in their own narrative; and understand that there are versions of themselves that are worthy of greatness, dragon slaying, and sex appeal.

This is a good thing, and it has been a long time coming. The rise of Indian and Chinese cinema, the Oscars to Denzel and Halle, the casual multiethnic makeup of TV shows like *House* and *Lost*, all those awards and plaudits that encourage men as straight as Sean Bean and Tom Hanks to play gay, and the fact that Will Smith is the biggest film star in the universe—these are all happy parts of this happy trend.

But why in all this have the atheists been left unrepresented? Where can we look for our myths, our stories, and our worldviews? Is it really as it appears, that in cinema we are the final frontier—and if so, why?

It seems to us that there are two reasons for this absence of atheist movies: one boring and prosaic, the other more interesting and perhaps more profound. Neither is going to change anytime soon, though.

The dull reason is that movies are mass market entertainments, while atheism remains a niche. With the average budget of a studio movie now in excess of $70 million, plus another $30 million in marketing costs, Hollywood isn't making movies for niche audiences. They are making them for the mob. And the mob globally—particularly in the world's biggest movie market, the United States, and in its fastest-growing ones, Asia and Latin America—is a religious one.

Good people, for sure, but as a mob, the same people who make it unthinkable that America could have an atheist for president (it became a landmark moment for atheists as well as black people when President Obama in his inaugural address included a reference to "non-believers") make it equally unthinkable that Disney is going to drop $100 million on a product with a core message that will ostracize and offend its biggest audiences.

It is this boring and inconvenient truth that has led the film adaptation of *Paradise Lost* to stall, that led *Contact* and *The Golden Compass* to have their atheistic, subversive hearts ripped out, and why *The Passion of the Christ* was the sensational success it was. It's a market thing, an externality to the creative process of storytelling—a constraint that for mainstream, big-budget movies is unlikely to go anywhere anytime soon.

More interesting, however, is the second reason, the internal, storytelling reason that no great atheistic movies have come down through the ages, not even smaller independently financed ones. Core to all religions are the principles of good storytelling. And core to all great stories are a religious moment.

All the religions have at their heart eternal, mythic stories of death and rebirth; of fathers and sons; of creation and destruction; of hope and despair, persecution and deliverance. They satisfy the most primal urges and needs of mankind's collective unconscious: the need for meaning and the need for relief from the fear of death. And those same needs are the ones that drive our need to tell and hear stories.

Religions exist, survive, and continue to flourish even in an era of abundance and science for one profound and irreducible reason: because they are the stories we need. As such, they are by definition the best, most resonant, and most powerful stories ever told. Thor and Odin, Hercules and Zeus, Adam and Eve, Shiva and Radha, Buddha sitting under the tree, Noah and the ark, even Joseph Smith and the astronaut angel in his back garden—damn good yarns, each and every one of them, and they influence us all, whether we believe in them or not.

And so this influence seeps into our celluloid tales. Like it or not, they inhabit our psyches, and they are the tropes and metaphors we reach for when trying to satisfy an audience. Even those of us for whom the old stories of religion are diminished in significance because we don't believe in their literal truth find ourselves somehow serving a categorical imperative of cinema to renew those stories, to supply new tellings of archetypical religious myths. The stories, somehow, will out.

This is not a new thought. Jung wrote of it first and best when he said, "You can take away a man's gods, but only to give him others in return." Storytellers of all stripes—from shamans and prophets, via mythologists like Joseph Campbell, to film directors and screenwriters; from Matthew, Mark, Luke, and John to Stephen, George, and Stanley—have understood this.

This demand that story fulfill metaphysical needs is manifest not only in the big decisions of what stories you tell but also of the small ones of how you tell them. The burning bush in *The Ten Commandments* is a cinematic antecedent to the floating plastic bag in *American Beauty*; Patrick Swayze's phantasmagorical sculptures in *Ghost* have their beginnings in the gospel story of the Holy Spirit taking over the tongues of the apostles; Buffy's ascent to heaven after she sacrifices herself to save her sister Dawn owes much to Joan of Arc; and Superman's descent to Earth from Krypton on a rocket ship made by his father, Jor-El, owes a lot to another mythological father who bequeathed his only son to save us.

These moments of supernatural epiphany, of pathetic fallacy, of cosmic intervention make for good cinema. And it is why, no matter how atheistic the creators might be, religious metaphors and imagery continue to pervade our work.

The same principle works in reverse: if religion has the best stories, it is because narrative, classically, requires a transformational moment, an epiphany, to function in the three-act structure beloved of film writers, and that epiphany tends to be moral in nature—from bad to good, or in a more modern system, from blindness to self-awareness or whatever—and that transformation may always need to have a religious quality. Love, the most common theme of the movies, tends to be found (normally by someone who up to that point has not believed in it) round about the end of the second act, in a moment more than a little reminiscent of the clouds breaking and a sonorous voice speaking out of the sun.

We found this ourselves in our current film, *The Infidel*, which started its life as a comedy with a message about the evils of the intolerance fostered by most religions. We didn't think when we started work on it that an atheist writer and producer, working with this premise, would make something that hinted at a belief in God. But the movie has gradually morphed, despite our best atheistic intentions, to include a key moment of divine inspiration and road-to-Damascus conversion. While it might not serve our didactic purpose, we cannot help but admit that it serves the needs of the film. Again, Jung got there first: "God approaches man in the form of symbols." What a smug know-it-all. (Jung, not God. Actually, the two of them can both step off.)

So, faced with the dual pressures of the religious film-going mass market and the internal traitor in our own storytelling psyches, what is a pair of atheist filmmakers to do? Is it even possible to come up with a mainstream, entertaining movie that deals with atheist themes, yet remains accessible and emotionally satisfying?

We think so—and in fact have come up with one. It's a romantic comedy. Atheist boy meets atheist girl. They are united in love and atheism. It's perfect. So perfect, in fact, that atheist boy starts to think it must have been divinely ordained, a match made in a heaven he finds himself believing in for the first time. Trouble is, atheist girl can't be with a believer . . . so she dumps him.

That's the first act, anyway. Now all we need to do is work out who plays the archangel Michael, who will come down from heaven to re-unite our young lovers. . . .

A Christmas Album

Simon Price

The next time some nit-picking ninny asks you why you celebrate Christmas and listen to Christmas songs when you don't even believe in God, please try to refrain from throttling him—it is, after all, the season of peace and goodwill to all men.

Instead, try politely pointing out, for starters, that the man who shaped the sound of our modern Christmas wasn't a Christian either. Harvey Philip Spector might not be everyone's idea of a devout and observant Jew, but he sure ain't no Santa Claus. It may feel counterintuitive that a misanthropic gun-wielding maniac should come to define the joy-to-the-world spirit of the festive season, but define it he did. Roping in the Ronettes, the Crystals, and other artists from his stable to record seasonal songs old and new, his trademark Wall of Sound augmented by a thousand jingle bells, Spector made the album—1963's *A Christmas Gift for You*—which, to this day, is the only Christmas record you really need (so much so that the only decent Christmas song released in recent-ish memory, "All I Want for Christmas Is You" by Mariah Carey, is a total Spector pastiche).

And the next time someone points out that the "true message of Christmas" has been lost among an orgy of consumerism and self-indulgence, not only are they absolutely correct, but—give or take a few centuries—'twas ever thus. For all Cliff Richard's hearty arm-swinging efforts to drag us back through the metaphorical church doors, the British have always been more inclined to celebrate in the same disgraceful, decadent, and hedonistic manner as their heathen ancestors.

Roy Wood, the terrifyingly hirsute leader of seventies glam rockers Wizzard (who, perhaps not coincidentally, bore an uncanny resemblance to a Dark Ages Druid, albeit one who had narrowly survived an explosion in a paint factory), understood this proud tradition when he prefaced his own Spectoresque classic, "I Wish It Could Be Christmas Everyday," with the wickedly cynical *ker-chinggg!* of a cash register. And so, several decades before her name became euphemistic rhyming slang for bowel movements, did purring jazz siren Eartha Kitt on "Santa Baby," even if her craving for a sable doesn't quite sit comfortably with modern sensibilities regarding animal rights.

What we're reaching toward, as you can see, is the rather wonderful truth that not being a Christian isn't necessarily a hindrance to enjoying Christmas—and the music that goes with it. On the contrary, it's arguably a very handy qualification.

Even if the singer of a Christmas tune genuinely is consumed by religious fervor, it's a relatively tiny obstacle. The pop critic, and any serious fan of music, learns pretty quickly that it's useful, nay essential—particularly in the case of the aforementioned Mr. Spector—to be able to separate the art (immortal and sublime) from the artist (all too fallible and human). When you've read the small print on as many album sleeve notes as I have, especially on recordings by soul singers, you soon become immune to God thankers and, after the slightest sigh of disappointment, turn a blind eye.

We all surely have a favorite Christmas carol, even if the rationalist voice in your head, if left unchecked, begins picking apart its metaphysical logic and its historical veracity. In the same way that one can gaze up in awe at a spectacular cathedral—or, for that matter, a fascist sports stadium—it ought to be okay to admit to getting tingles when you hear "Once in Royal David's City" or "O Come, All Ye Faithful" without feeling like a sellout.

Even for secular pop lyricists, the language of religion provides a handy set of references, a cultural shorthand that's there to be plundered, from the most overwrought crucifixion metaphor to the most innocuous use of the verb "to pray" and the least literal exclamation of "My God!" If atheist listeners tried to rigorously excise that stuff from their record collection, they'd be left with a handful of instrumentals. And, most likely, no friends.

Another reason that's often given, by believers and heathens alike, for switching off the radio as Christmas approaches is that the songs tend toward the sickly and the sentimental. While there is undoubtedly a lot of truth in that—one need only consider ye olde standards like "White Christmas" by Bing Crosby or "The Christmas Song" by Nat King Cole for proof—it's by no means the whole story.

Christmas provides an irresistible backdrop for the mini-drama of the pop song. Romance is, as songwriters instinctively know, that bit more romantic when conducted under cheap Chamber of Commerce fairy lights to the distant screams of Noddy Holder, and tragedy, as the scriptwriters of *EastEnders* are aware when concocting their fiendish doof-doof finale, that bit more tragic.

In the words of Jerome K. Jerome, "Nothing satisfies us on Christmas Eve but to hear each other tell authentic anecdotes about spectres. It is a genial, festive season, and we love to muse upon graves, and dead bodies, and murder, and blood."

Why else would so many people feel inspired to throw their arms around each other's shoulders for an emotional sing-along to the

Pogues' "Fairytale of New York," in which a pair of drunken old derelicts, wishing away what's left of their sorry lives from the discomfort of a police cell, serenade each other with the unforgettable words, "You scumbag, you maggot, you cheap lousy faggot, 'Happy Christmas' your arse, I pray God it's our last"?

Heartbreak and Christmas go together like plum pudding and brandy sauce. If you want evidence, just listen to Darlene Love's tear-jerking "Christmas (Baby Please Come Home)," this writer's favorite Christmas song of all time. Other examples of tears among the tinsel include Elvis' "Blue Christmas," Mud's distinctly Presleyish "Lonely This Christmas," and, of course, Wham!'s immortal "Last Christmas," and even "Christmas Time" by the Darkness (although the Hawkins brothers sabotaged their song's poignancy somewhat with the innuendo-riddled chorus "Don't let the bells end . . . just let them ring in peace").

Happiness pure and simple, on the other hand, is banal and boring. This is why nobody remembers the eighties efforts by Shakin' Stevens and Gary Glitter with much fondness (the latter, admittedly, is also scuppered by the sinister edge the line "You'll never guess what you've got from me" has acquired in hindsight).

"Do They Know It's Christmas?" by Band Aid, the biggest-selling Christmas single of all time, is unsurprisingly (given the tight turnaround in which it was written and recorded) an ugly and jumbled thing, its titular rhetorical question providing successive generations of smart alecks with the opportunity to reply, "Given that 66.5 percent of Ethiopians are Christian, of course they do" (not to mention the chance to point out that of course there will be snow in Africa this Christmastime, at the peak of Mount Kilimanjaro, which has a 365-day covering, at least until 2050, when scientists predict it will all have melted).

In its defense, that song's most histrionic moment, the fist-clenching, neck-bulging contribution from Mr. Bono Vox, has given us one glorious addition to the English language. Friends of mine have become accustomed, every December, to receiving a card wishing them "clanging chimes of doom." It's a joke that never gets old, or at least, if it does, they're too nice to tell me so.

Band Aid wasn't the first attempt to graft a humanitarian message onto Christmas. John Lennon used the season as a platform for pacifism with "Happy Xmas (War Is Over)," as did Jona Lewie with "Stop the Cavalry," in which the pub rocker adopts the persona of the universal soldier, from the trenches of the Somme to the nuclear fallout shelters of the Cold War, while an old-fashioned colliery band parps away in the background. The thinking behind these songs is that Christmas encourages us to consider the plight of others for a few fleeting days before we get back to bashing each other over the head with large pieces of metal. (The folly of which would become apparent to anyone who actually watched the December 25 episode of the aforementioned *EastEnders*.)

They do say Christmas brings everyone together. This power has never been demonstrated as starkly as in the video to "Little Drummer Boy," in which unlikely "neighbors" David Bowie, then at the height of his coked-up Berlin phase, and sweet old Bing Crosby, three weeks away from death and wearing a golf sweater, start pa-rum-pa-pum-pumming away together like bosom buddies, in what must surely go down in pop history as the most surreal duet of all time.

Even a staunch bah-humbug stance can sometimes be overcome by the general festive mood. In the Waitresses' cult classic "Christmas Wrapping," the protagonist's sour post-punk cynicism is melted unexpectedly by a chance encounter in the frozen-turkey aisle of the late-night grocery. And bonhomie is always to be welcomed, wherever it comes from, even if there's an ulterior motive: when you hear that sly old dog Dean Martin singing "Let It Snow," for example, you just know it's only so that he can cop a feel under the mistletoe.

So this year, if anyone accuses you of hypocrisy for engaging with the Christmas spirit, don't give him too hard a time. Crack open the brandy, crank up the Crystals, and remove your hands from around his neck.

Rockin' Around the Christmas Tree

NATALIE HAYNES

Blitzes on Christmas by the irreligious are the stuff of tabloid headlines every year—according to them, atheists are always banning Nativity scenes, Santa, Rudolph, and a Christmas Day showing of *The Great Escape* on BBC2. Which is kind of strange, because most godless people I know are in favor of Christmas and wouldn't dream of censoring a glass of mulled wine and a viewing of *The Bishop's Wife*, partly because they aren't rude enough to tell other people what they can and can't enjoy, and mostly because they have already rented the Cary Grant movie, made the popcorn, and settled down on the sofa.

To the non-Christian, Christmas represents something very different from its official job description, but what it represents certainly isn't trivial. The values behind Christmas—I mean the ones that should, in my view, underpin Christianity but so often seem to get lost—are ones I think many non-Christians share. I'm not crazy about the baby, the shepherds, the kings, and the virgin birth, but loving one another, forgiveness, generosity? Most of us would agree that the world could do with a bit more of those.

So I, at least, choose to keep Christmas with gusto every year—the tree, the demented cooking, the presents, the surprises. Dammit, I make my own crackers. I am, frankly, Christmas hardcore. And the reason for that is my mother, who has always kept Christmas like a post-ghost Scrooge.

Throughout my childhood, I remember Christmas being as it should be: magical. Of course we put mince pies and carrots out on Christmas Eve; we made Advent calendars and cards. For my mother, Christmas was one long distracting-us-from-rainy-days craft project. Pritt Stick, felt-tip pens, and colored cardstock were our metier. I don't think we had a window without a skewed paper snowflake adorning it. She's had her home recarpeted twice since then, and there is still glitter lurking.

The preparations for Christmas are, I think, what I like about it most. Don't get me wrong—I'm fond of a present—but the buildup to something is usually more fun than the thing itself. As a child, I always loved the process of buying a Christmas tree, getting it home, and decorating it. Once it was done, I still loved it, but the exciting bit was over. Every year my mother would suggest buying an artificial tree, to save us the trauma of getting our tree too early and seeing its needles fall sadly to the floor (those have also, I believe, survived the recarpeting) by December 20. Or the opposing fear: that by the time we made it to the garden center, all the good trees would be gone, and we would be left with a gnarled twig that could barely support a single bauble. Either way, obviously, Christmas would be ruined.

Every year we would wail that a real tree made it Christmas. That the house wouldn't smell of Christmas without one. That Christmas wasn't Christmas if no one got stuck in the corner trying to pull the netting off a Nordman fir without losing an eye. My mother would mutter something about a scented candle and then concede defeat.

You know those trees people have where the baubles are beautiful, two-tone, and matching, like the ones in a fancy department store?

Well, ours was, and is, nothing at all like that. At no point could you mistake it for one decorated in nuanced fashion by a window dresser with ambitions. Rather, it looks like what it is: a chaotic mass of every year we've been alive, distilled into wood, glitter, and foil. Every year my mother asks me if she can get rid of Zebedee—an aged, faded bauble-representation of the *Magic Roundabout*'s star. And every year I look at her like she is mentally deficient. The very idea of binning Zebedee seems to me roughly on a par with kicking a puppy.

And once the tree's up, of course, there's the cooking. I have been cooking Christmas dinner for as long as I can remember, certainly since I became vegetarian at the age of twelve. People always ask sorrowfully if I don't feel I'm missing out on the turkey and its trimmings, but since I haven't the least recollection of the taste of it, I can honestly say I don't. Plus I can cook anything I like for Christmas—a choux-and-cranberry ring, a leek-and-chestnut crumble—without fear that someone will moan that they aren't getting what they want. And I can line it with all the roast potatoes I can eat—a considerable number—which are the objective highlight of any meal.

It seems to me, in my rather biased view of things, that the turkey is the most miserable bit of Christmas. Even if you like it, you have to be awake before dawn to cook the bloody thing. And how many magazine articles are devoted to ways in which you can cook a turkey to prevent it becoming dry? It seems demented to me: why not cook something that doesn't, as its default setting, taste of nothing? And, by the way, why not cook something that means you can have a lie-in? It's Christmas, after all. Spoil yourself.

Again, I often enjoy the run-up to Christmas, foodwise, more than the day itself. Call me greedy, but by Christmas Day I have usually eaten all the mince pies I'm going to for that year. This is because I cook like crazy in the run-up to Christmas—another hangover from my childhood. If you come round mine any time in December, you're likely to be offered a festive bun, or cake. This is not, you understand, because I am trying to achieve some kind of domestic deity status. When I was a child, my baking obsession was down to eagerness; now it's mostly work avoidance. Either way, you still get a cookie.

Christmas plays a huge role in our cultural lives, from the endless repeats of *It's A Wonderful Life* to the endless remakes of *A Christmas Carol*. Again, it's no less true for the irreligious—I cry every time I watch *It's a Wonderful Life*. You don't have to believe in angels to watch the movie, any more than you need to believe in dragons to read *Lord of the Rings*. Christmas has surely inspired some of the most brilliant television, films, and books that we have ever created, and many of them date back to my childhood.

I can't imagine the run-up to Christmas without watching *Box of Delights*, the BBC's adaptation of John Masefield's classic children's story. I realize that by confessing this to you, I run the risk of being seen as an emotional simpleton who can't accept her age, but I don't really care. I behave with impeccable adultness most of the time. Even in December, I am happy to discuss the complexities of Aristotle's *Nicomachean Ethics*, if you really want to. But once we've done that, I intend to slide into the world of magical realism for a week or two, because it's been a long year.

And, by the way, I'm not even close to joking when I say that I think the Muppets' version of *A Christmas Carol*, released in 1992, when I was already, objectively, old enough to know better, is the finest adaptation of the book I've ever seen. And it's Michael Caine's finest film role to date, playing Scrooge with genuine pathos and passion. One of my favorite things about *A Christmas Carol* is its atheism: there are supernatural ghosts, sure, but the Marleys dwell in an afterlife that has as much in common with Greek myth as limbo. There's no suggestion anywhere that Christianity is the key to Christmas, no trip to church for Scrooge on Christmas morning. Rather, kindness and tolerance, generosity and hedonism are what is required to give Christmas its due.

I hope that's what people find when they spend Christmas with me. It's certainly what I grew up believing Christmas should be—my mother was a far bigger fan of Dickens than Matthew, Mark, Luke, or John. And like the haunted Scrooge, I try to keep the spirit of Christmas, his Christmas—gifts, food, excess, fun, and love—every day. And I say bah, and quite possibly also humbug, to those who disagree.

God Isn't Real

Robbie Fulks

One August afternoon twelve years ago, as I sat daydreaming on my suburban back porch, a churchy three-quarter Appalachian melody came humming into my head. I picked up a composition book and jotted down some words:

God Isn't Real

A world filled with wonder, a cold fathomless sky
A man's life so meager, he can but wonder why
He cries out to heaven, its truth to reveal
The answer, only silence: for God isn't real.
Go ask the starving millions under Stalin's cruel reign
Go ask the child with cancer who eases her pain
Then go to your churches, if that's how you feel,
But don't ask me to follow, for God isn't real.
He forms in his image a weak and foolish man
Speaks to him in symbols that few understand
For a life of devotion, the death-blow he deals

We'd owe him only hatred—but God isn't real.
Go tell the executioner of the power he can't defy
Go tell his hapless victim of the mercy on high
Then go to your churches, go beg pray and kneel
But don't ask me to follow, for God isn't real.
No, no matter how He should be . . .
God isn't real.

Makes a statement, I thought, and set down the pencil. A statement, assuredly, doesn't equal a song, and not everyone who agrees with *x* may dance to it, while others favoring *y* . . . well, I was shortly to find out about the *y* people. The most obvious limitation of overtly literal lyrics like the above is an effect of slightness. Epic narratives, moody word paintings, and love songs, by comparison, show their weight in labor and confession. A flexibility in its meanings is useful to a song, which stands to be performed or experienced, pardon the saying, God knows how many times. Still, the polemic has its appeal—easier to frame and finish, and more open to acidity and playfulness. So I counted "God Isn't Real" a keeper, and a week later knocked out a demo and mailed it to Affiliated Publishers, Inc., the generically named Nashville concern for which I was a staff songwriter, and who contractually owned copyrights on all I made up.

As expected, I received no response. They were over me by this time. Since being hired I had written them a good number of empowered-women-in-or-out-of-love efforts, following the mid-1990s commercial template. But mixed with these were so many oddments—talk-radio themes, fiddle tunes, spirituals from imaginary Depression-era musicals, barroom weepers, amused little riffs on pig meat and pocket pool—that they finally concluded it was more profitable to turn tastefully away than to sift. If I thought I might prod these peaceable Tennesseeans alert with my brief against the Bearded One, I was in error. The blond policeman's wife who typed lyric sheets and sent titles to ASCAP and BMI for copyright registration was, like the rest at API, an observant and faithful church-goer who was nonetheless, in matters of business, consistently unsupernatural.

So my musical infant might have died quietly, with its hundred-odd siblings, at API's doorstep. But I also had a sideline as a performer, and before long I was trying out what I assumed was country music's first atheist ballad on my ragged little audience of honky-tonk traditionalists and urban hipsters. Their response commingled chatty indifference—put a low-key waltz about theology before a Saturday night dancehall crowd yourself if you don't believe me—with sharp feedback at the margins.

Two old ladies in Oklahoma City walked out on me, loudly, with a skreek of their folding chairs and grim-set faces. At a club in Hoboken, a woman standing alone by the back wall, tall and fine-boned and dressed like a jet-setting David Lean heroine, got switched on by the lyrics; at my next date in that town she returned with a rich-rouged gaggle of disbeliever girlfriends to form a little cheering section—godless girls gone wild. In Houston an older married couple waited in line to tell me tremulously that I was a messenger of evil. At Second City, the improv comedy institution in Chicago, the tune went over like gangbusters and soon became a staple of my Christmas shows there.

In 1998 I recorded "God Isn't Real." The studio, once owned by the country singer Ronnie Milsap, was back in Nashville. To up the song's authenticity, I called in a few A-list old-timers to play. But as tape rolled, I found myself stricken by cowardice, unable to intone the sacrilegious phrases. The great and gentlemanly John Hughey was pulling heart-wrenching swells from the pedal steel, as he had on a hundred Conway Twitty and Vince Gill songs, while I was lamely burbling from the vocal booth: "A hm full of hm-hm / A cold hm-hm hm-hm . . ." Later, after my vocal was done and there was no disguising it, another name picker came in to overdub. Emerging from the booth, he approached me, mandolin in hand. "About this song," he said, and his voice fell to a whisper, "I actually kind of agree with it."

Did I, though? Some folks have asked in the years since. I'm surprised that that set of lyrics leaves much interpretive room. But I've read the suggestion that the song is anti-establishment, not anti-theist, while another listener detected devious satire—a believer's mockery of coarse-witted atheism. Though deviousness isn't my thing, I do confess to a dalliance with belief.

At fourteen, impelled by curiosity, rural loneliness, and a desire to know which group I stood with, if any, I resolved to read the Bible from front to back. I had inherited a small-town Protestantism that, more and more, felt unexamined and ill-defined. Aside from occasional services and before-meals prayers, its ritual demands and doctrinal content were about zero. Much of what my family valued—social justice, honest work, some capacity for personal humility, and quiet inward focus—reflected a Christian attitude without requiring any particular stance on ancient miracle working in the Middle East.

As I read the book, a mealy King James distillate for "today's teens," at the rate of a few chapters a night, I began trying out a few of the churches in the little town a few miles from our North Carolina farm. The Baptists opposite the bank on Main Street had a clarion-voiced preacher who looked a little trashy to me, with anchorman hair and electric blue sport jacket. Across town, in a white clapboard structure on a scraggy parcel of land inclining down from the street, sat the evangelical black church. The service there went on for hours, with intense singing, linked hands, a hug-your-neighbor interlude, and, after the sermon, a home-cooked lunch in the grass outside with yet more singing. Uplifting, and welcoming—but, sitting among them in my Saxon pallor, I might as well have worn a pith helmet marked ANTHROPOLOGIST.

The Methodists were a better fit. Their pastor was a gentle, white-haired man with the inapt name of Rouse. The man exuded a reassuring lack of passion. One of his deacons was a stocky man in his mid-twenties, whose full but neatly tended facial hair was 1977 fashionese for "I totally get where you youngsters are coming from." With two other teenagers, I joined his Wednesday night Bible study. These classes, it turned out, were not laid-back, Me Decade–style informal group sessions, but stupefying accounts of tribal clashes and the plodding movements of populations in the Canaan region some 1,977 years previously. By the third week my two classmates had disappeared, and as a consequence, the tone of Bible study night changed straightaway to more of a laid-back dialogue. Sitting together on the side steps of the handsome old church, my with-it tutor and I rambled freely. He was like me—not quite fitted to Hickville. One night he handed me a book, a slender, modern-English vade mecum that might, he suggested, be used as a thought-provoking scriptural adjunct in my search for higher purpose.

Jonathan Livingston Seagull did tie in with the Bible, loosely. (Smoking a joint helped clarify the connection a little.) It counseled sensual renunciation: "Don't believe what your eyes are telling you . . . look with your understanding" echoed Corinthians' "For we walk by faith not by sight." Faith and resurrection were encoded in the theme of flight, by which Jonathan is determined to lift himself from his earthly bounds and his transcendentally tone-deaf flock, who mainly think about eating. A gladsome vibe of seventies self-empowerment suffused the book. "We can lift ourselves out of ignorance, we can find ourselves as creatures of excellence and intelligence and skill. We can be free! We can learn to fly!"

These motivational "cans," especially next to the scary and admonitory Old Testament, looked a lot like cant. The Bible's a messy book, but all its ambiguities, contradictions, numerological digressions, magic, clan trivia, anachronistic injunctions, bad science, and bloodletting make for a much better read than a streamlined modern book delivering a vaguely Jesus-y uplift. Oddly, the Bible's flawed coherence and cohesion can have the effect of burnishing its authority, its flesh-tingling lifelikeness.

Sadly, these same qualities make the Bible impractical as a guide for living, or so I was coming to think as I neared Revelations. An eye for an eye, or turn the other cheek? Love thy neighbor, but sacrifice thy son? Wisdom—a joy or a misery? Make up your corporate mind. Speaking of which, who wrote all this? And how many generations of translation stood between me and whoever that was? The Christians around me were as in the dark as I was. Tellingly, they didn't follow half the teachings of their holy book: their women wore pants, for instance, and wouldn't consider abandoning their families to follow gentle teacher-seers into the desert. They picked which church to attend, and by extension which teachings to focus on, according to their own temperaments and cultural makeup. I couldn't reconcile the Bible's status as scripture with the mix-and-match approach it made necessary, or reconcile the actuality of a million creeds with the idea of a single Creator, a unified and unimpeachable truth. This wore away my effort at faith, which over the years was further eroded by the evidence problem, and the problem of suffering laid out in "God Isn't Real."

Turning away from God definitely did not lay to rest any of the elusive objects that brought me to seek him—"what is true" and "how best to live" and so on. (Possibly a stereotype held by believers about their opposite numbers is that, having cleared our desks of the slipperiest of humankind's eternal questions, we live out the rest of our days in shells of glib materialistic certainty.) Country songwriting, though it may be a tool of sub-Aquinas subtlety, makes for a pretty mean existential chisel in the hands of its best practitioners, to whose ranks I aspired as a mature man. Twenty-one years after leaving Methodist Bible study, I was back in the Carolinas, now as a professional troubadour. Along with my band, I was performing "God Isn't Real" night after night, promoting my new record and alienating the devout across multiple markets.

My friend Dan was the drummer. A native South Side Chicagoan, he had a stolid demeanor and a near-chromosomal bias toward the practical and the provable. We were tooling along a state highway near Raleigh and talking about the previous night's reception, which had verged on fists joining faces. "I don't get why people get so bent out of shape over that song," he said.

"You don't?" I said, surprised. "I do. People take their religion seriously. And this is North Carolina, for God's sake."

He shook his head skeptically. "But we're not playing in churches. And I don't see where's the offense in what you're singing. It's a point of view. Nobody's forced to agree with it."

"Well . . . that's a sensible statement about an inflammatory song about an emotional subject. I'm not a believer, but I'm much more in line with the one guy who's mad at me for cutting down God than the ten guys who're bored with the whole idea and yelling for a louder song. I mean, the question of what happens after you die, the mystery of who if anyone is minding the cosmological store . . . those are important questions, right? I can't think of anything more important than that."

He studied the road ahead for a few moments, then shook his head once more and finally. "I guess I never worry about that stuff."

The vital contingent of Americans to whom things like string theory, Keynesian economics, and Baptist theology are just three of the many faces of "that stuff' represent a healthy, commonsense strain in our national life. The highly touted culture war is just never going to take hold here in the States. Too many people are too content to raise kids, make a dollar, grill a burger, go to the show.

I'm not one of them. I honestly think "He should be" real, as my song's coda says, and while I think we should be able to laugh with 100 percent lung capacity at any subject, I don't take the evident absence of a celestial Creator as cause to crow. Take away God, and our basic problems—what to make of our lives and desires, how to address our hardwired need for a kind of higher narrative order—suddenly become much more vexing. And our shelves fill with gaseous books about bellyaching birds.

I began with a cocky diatribe. Let me end with a call for a little epistemological humility. We skeptics would do well to emulate that titan among skeptics, H. L. Mencken, whose pro forma printed response to all letters of disagreement was "You may be right." Humans are higher animals only relative to such as dogs and turkeys, and it's possible, as has been noted, that our brains as presently evolved are no better equipped for scanning the cosmic fabric than a poodle's (or mine, actually) is for mastering math. This sobering possibility should ground all metaphysical editorializing.

That's why, besides tiring of the rigid lyrics after a couple hundred performances, I stopped doing "God Isn't Real." Einstein spoke with poetry and power of "this huge world, which exists independently of us human beings and which stands before us like a great, eternal riddle." On this ground, I believe thoughtful believers and atheists alike can stand in real, if fragile, alliance.

The Godless Concerts

ROBIN INCE

"Ladies and gentleman, that was Jarvis Cocker. Now please welcome Professor Richard Dawkins!" And the Hammersmith Apollo continued to go wild.

In my life of low achievement, my high remains putting on a show where Jarvis Cocker sang "I Believe in Father Christmas" before Richard Dawkins came on to talk of the size of the universe and ape behavior, which was followed by an interpretative dance to "I Can't Live if Living Is Without You" by *Pan's Person* Joanna Neary.

It all happened because of a feud on a regional television debate show. I had been asked to appear on *London Talking* as a mouthy stand-up comedian atheist. The researchers had seen a debunking of intelligent design titled "Magic Man Done It" on the Internet and thought I could add a little levity to their TV debate on the question "Is Britain becoming more secular?" In the end, rather than levity, I injected a sort of crazed rage that might normally be seen from a man dressed in hessian with carrier bags for shoes. This is the risk of agreeing to enter the fatuous terrain of TV debate shows. They are, after moments of extremism, delivered as succinctly and volubly as possible, with a neat wrap-up as they take you into the ad break. No member of the TV audience should gain anything but a reupholstering of his preconceived notions.

When I arrived at the studio I had an inkling things might go awry. The debate had transformed from "Is Britain becoming more secular?" to the far more tabloid-warming "Who is taking the Christ out of Christmas?"—the favorite of lazy journalists who have a pathological fear of facts.

I had been informed that the debate would involve religious heavyweights. Sadly, this did not mean Archbishop Rowan Williams or a papal emissary. Instead, the religious goodness of Britain was represented by radio shock jock Nick Ferrari and by Vanessa Feltz, people who form opinions for financial gain and celebrity. Also within this televisual general synod sat Stephen Green, leader of Christian Voice, a marginal but caterwauling gang who are as representative of British Christians as a black and white minstrel is representative of multiculturalism.

Within seconds of the debate beginning, I was puce and itchy. Debunked story after debunked story was dragged out again and spouted as truth.

When they finally got to me there was no time for facetious levity, and in the few allotted moments I had, I thundered through evidence. The other guests and audience members were not fans of evidence or facts; they had heard their hearsay and that was that. Despite the number of Christmas cards in shops saying "Happy Christmas," you could not buy a card saying "Happy Christmas." Despite the Christmas lights in town upon town, no lights could be seen. Despite the carol services in every church in the land, there were no carol concerts. They had plucked out their own eyes, and now berated others for what they could not see.

I also made it clear that, as the antsy atheist rep, I was not against celebrating Christmas; in fact, I rather enjoyed this time for contemplation, ginger wine, and gift disappointment. Each time I suggested this, Stephen Green would smugly point out that he knew I would like Christmas banned. It didn't matter that I said that it wasn't true; God had been in his head and told Stephen I was lying.

Some weeks later, I had finally calmed down and started to sleep again. It was then that I decided I would prove to Green that atheists could enjoy a Christmas celebration as much as the devout. Rather than celebrating mythical wise men, angels, and the mythical messiah, we could celebrate rational thought and scientific thinking. My years of Carl Sagan and Richard Feynman worship would come to fruition.

Fortunately, in the next few days, I bumped into Richard Dawkins while appearing on the *Richard and Judy Show.* He discussed the importance of Charles Darwin while I showed funny clips from the Internet of men being peed on by incontinent puppies. After a brief conversation on the way to his cab, Richard Dawkins looked like he would be the first of the guests on the rational Christmas show. I soon found out that it was not going to be difficult to gather together a group of musicians, scientists, and comedians to sing the praises of science. Charlatan destructor Ben Goldacre, physics popularizer Simon Singh (whose participation reminded me I really must finally read that copy of *Fermat's Last Theorem* that had been perched on my bookshelf of good intentions), *Jerry Springer: The Opera* librettist Stewart Lee, and singer and giant-killer-crab lover Robyn Hitchcock all said yes.

My early inklings of what the evening should be had me in a vaguely facetious frame of mind. Stephen Green had worked hard to campaign against *Jerry Springer: The Opera*. His followers stood on pavements outside theaters singing hymns with the ugly accompaniment of portable electric organs, while passing out leaflets feeding misinformation and downright lies about the work. Green had boasted that, thanks to his work and lies, *Jerry Springer: The Opera* would never be performed again, and so I felt that maybe the night should be an outing for a few of the juicier arias.

After initial excitement, I realized that I would be falling into a Stephen Green trap. If we aired the arias, we would soon be discounted as no more than smug atheists rubbing Christians' noses in nappies and well-sung swear words. The night had to be joyous. Lazy hacks (of which there is no shortage) are quicker to damn a popular atheist than a religious fundamentalist. Check the Melanie Philips *Daily Mail* column of November 26, 2007, for proof: "The Real Nutters Are the Fanatics Who Despise Religious Belief." Even real journalists write similar things. If atheists/agnostics are ever going to win over the loosely religious, then it won't be through constantly telling them they are half-wits for clinging to their deities, popes, and mullahs. As Mark Twain wrote, "It is easier to fool people than to convince them that they have been fooled."

As the night took shape, I realized that it should take the form of fractured variety versions of the Royal Institute lectures, as if Niels Bohr were sandwiched between Jimmy Tarbuck and the Swingles Singers.

With no publicity save for a few plugs on science sites and skeptic websites, the first night sold out at the 550-seat Bloomsbury Theatre. Everyone seemed keen on appearing at a second night, and that sold out too. Then the madness of the megalomaniac took over. Surely we should put on a final night at the 3,500-seat Hammersmith Apollo, home to Billy Connolly and the death of Ziggy Stardust.

There was no skimping on expenditure—we would have a grand star curtain, a big screen to project Carl Sagan's Pale Blue Dot onto, a twenty-four-piece orchestra, plus some secret star guests should the tickets not sell on famed evolutionary biologists alone.

The first night was longer than *Hamlet* but slightly shorter than *Once Upon a Time in America*. The theater manager said he had never had so many professors collecting tickets from the box office, but it wasn't just geneticists and cosmologists. There were teenage Goths, twelve-year-olds in cast-off woolens, and a few twentysomethings in RICHARD DAWKINS IS GOD T-shirts. "Oh, dear, that means I don't exist," said the worried professor.

The one reference to wave/particle duality got a big laugh and a round of applause, which is not a regular occurrence in comedy clubs. Tim Minchin, barefooted Australian comedian and rock god, performed a nine-minute beat poem about woolly thinking; Simon Singh reminded us of the size of the universe via the rewritten lyrics of a Katie Melua song; Ben Goldacre attacked vitamin salesman and AIDS denier Matthias Rath in a somber and compelling ten minutes; and Stewart Lee told us that he now believes in God because of Richard Dawkins, saying, "When I look at something as intelligent and intricate as Richard Dawkins, I do not believe it could have happened by chance."

It was a long night. As the Anglican Nine Lessons and Carols involves exactly that, I felt we had to have nine songs and nine spoken-word pieces, but comedians are not fans of brevity in front of a delightful audience. We missed closing time—not bad, considering we live in the age of twenty-four-hour drinking.

The second Bloomsbury show was equally long, and I realized I'd missed a trick by not selling godless comfy cushions for those in the harder seats.

Hammersmith merely became surreal: it was Christmas night with the stars, just that there was a higher quota of stars with Ph.D.'s than usual. It was one of the few shows at the Hammersmith Apollo that should have had a reading list given out at the end. I hope more people found the DVD of Carl Sagan's *Cosmos* under their Christmas trees because of it. There was no attacking of religion as a whole, just a few slights at some of the more unpleasant fundamentalists. Mostly it was jokes about astronomers with golden noses, the web-making abilities of spiders, and a song titled "Me and You, a Monkey, a Teddy, a Deaf Kid, and a Shoe" sung by mentally confused surf rocker Colin Watson. Sadly, there just aren't that many carols about monkeys, teddies, deaf kids, and shoes.

One Catholic journal was openly disappointed that there weren't any attacks on their deity. Maybe we'll change tack, but why knock the nonexistent when you can praise what is really there? I had plans for rational Ramadan, but it seems some theaters are a little anxious about that one.

EVENTS

Isn't it enough to see that a garden is beautiful without having to believe that there are fairies at the bottom of it too?

—Douglas Adams

God Trumps

CHRISTINA MARTIN

Writing is a strange occupation. When you try really hard to think of ideas, you come up totally blank, and when you're sitting around on your backside procrastinating, inspiration strikes. It really doesn't do much to encourage any kind of work ethic, I can tell you.

Here's a case in point. I was having a very unproductive day. None of the ideas I was working on seemed to have legs. As always in these situations, I abandoned what I was doing, made a cup of tea, and went online. I checked my Hotmail account, updated my Facebook status, had a quick glance at Google News . . . and that's when I saw the headline that started it all off:

> Gay Rights Don't Trump Christian Rights Say Christian Rights Group

What a ridiculous statement, I thought. *As though people's rights and beliefs can be reduced to a metaphorical game of Top Trumps . . . Actually, now you come to mention it, that would be quite funny!*

Fast-forward to a couple of months later (November 2008) and God Trumps—a card game that lists and ranks the habits and foibles of the major belief systems, and some minor ones—had been written by me, illustrated by *Guardian* cartoonist Martin Rowson, and published by *New Humanist* magazine.

I filed the issue away in a cupboard along with all my other clippings and thought no more about it.

Until, that is, *New Humanist* posted it on its website in February 2009 and, to quote the cool kids, it went viral. It was being forwarded, favorited, blogged about, and posted on discussion forums at an incredible rate. More than 56,000 people visited the *New Humanist* website the first day God Trumps went online—the average daily visitor count is about 3,000. At the time of writing, God Trumps has had more than 120,000 hits and is still the most viewed article in *New Humanist*'s online history.

This massive online proliferation obviously led to a lot of reaction and debate, and while most people, including religious folk, appreciated the God Trumps for what they are—a comment on the sheer number of belief systems that exist (they can't all be right!) and the bickering that goes on between them—some people inevitably took them a tad too seriously.

I saw forums containing pages and pages of sober theological debate sparked by the trumps, which is weird when you consider, for example, that the Catholic card has "Popemobile" named as the "Weapon of Choice." It's quite hard to take that seriously, I would have thought, much less have an impassioned debate about it!

There were also conspiracy theories flying about as to why certain religions were excluded. For example, there was no Mormon card in the first set, and one blogger wrote, in delightfully conspiratorial tones, "I think it's more likely that the Mormons were left out on purpose. Yes, folks, the Mormons are that powerful." Or perhaps we just didn't have room? And besides, who's scared of Mormons?

Anyway, we couldn't have been that worried about bad reactions. We had the guts to include the famously shady and (allegedly) litigious Scientologists. I'm saying "allegedly" partly as a joke and partly because they really are famously litigious (allegedly). This could go on forever, so I'll stop now.

Returning to people's reactions, there was also the person who actually thought the trumps were a genuine set of cards to be used in the process of choosing a religion. No, really. Although maybe that's not as ridiculous as it sounds—play the game, and whichever faith wins, you take it on. That's no dafter than just following whatever was foisted on you by your parents, I suppose.

One of my personal favorites out of all the reactions was from the agnostics. They were discussing the trumps on their forum and got really annoyed at their card, which gives them a noncommittal 5/10 in every category. They retaliated by mocking up a sarcastic "atheist" card to get us back, thus replicating in real life the inter-belief bickering that the trumps attempt to lampoon. (And just to digress for a brief moment, what on earth does one discuss all day on an agnostic forum—"I'm still not sure," "Me neither, I remain completely undecided"? Most odd.)

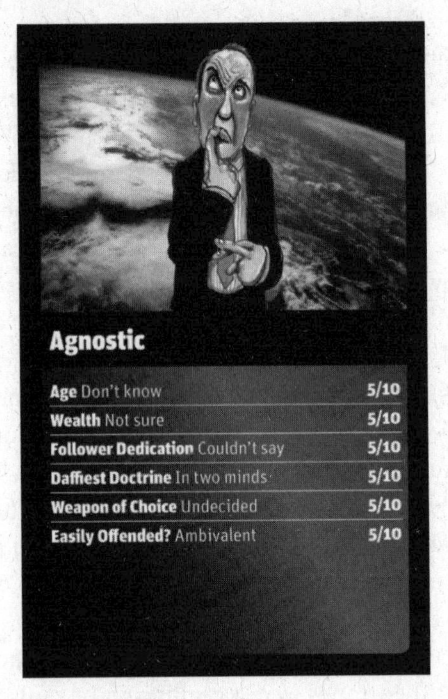

But returning to the topic at hand, there were quite a lot of people mocking up their own cards. Some of these people were motivated by a desire to see their faith included in the fun and games, others by a sense of dissatisfaction with the original version. One blogger, for example, mocked up an alternative Muslim card, because some considered our version to have been a cop-out. Which leads me neatly to the biggest reaction we received—and yes, it was about Islam. But not in the way you would expect.

Telegraph blogger and editor of the *Catholic Herald* Damien Thompson wrote an article titled "Humanist Attack on Religions Chickens Out of Criticising Islam" wherein he accused the piece of not tackling Islam properly for fear of reprisals. Which begs the obvious question: if we were scared of reprisals, why did we include them at all?

I should explain that the Islam card does not have conventional categories like the others. It is instead blanked out and designated as the ultimate trump card on the grounds that nobody is allowed to joke about it.

The irony went over Mr. Thompson's head, and he used it as an opportunity to knock us for it. In seriousness, all of this surprised me greatly, because when I wrote the piece, although I had expected the Islam card to cause some degree of controversy, I had thought it would be for quite the opposite reason.

The card employs a deliberate (and I would say possibly controversial) stereotype that Muslims are humorless, that they react badly to jokes made at their expense, that they burn effigies over trifling things like cartoons and therefore cannot be treated with the same good-humored approach as everyone else.

In my view I was neither targeting them nor copping out of dealing with them. The fact is that the cards used stereotypes about each faith to comic effect. This was the Muslim stereotype, and at the risk of offending them, as a comedy writer, I had to run with it, because I have more interest in matching the right joke to the right person than in being either politically correct or politically motivated.

Their positioning as the unbeatable trump card also came down to the very mundane fact that I used to play Top Trumps Ghouls and Ghosts as a kid. This game contained an ultimate trump card that had 100 points for every category and was pretty much invincible. As

I was modeling my piece on the actual Top Trumps I used to play, I had planned in advance to replicate this feature in my version, for old time's sake.

What's more, Islam wasn't even the front-runner for the title of ultimate trump. I had initially considered the pope as a contender because of his papal infallibility—that must come in extremely handy for someone so often in the wrong—but after some thought I came to the conclusion that, certainly in the current climate, it was more suited to Islam.

My real thought process couldn't have been further away from the false motives that were being assigned to me—cowardice, hedging my bets, fudging the issue. But what surprised me even more than this gross misunderstanding of my methods was the total lack of humor displayed by not only Mr. Thompson but the many people who shared his view.

The gag on its own was patently obvious, and was further elucidated by some very clear signposting, such as the illustration—a mad mullah—and the very pointed wording: "Well done to the extremist section of this faith for making it impossible to have any kind of reasoned debate or even a good-humored laugh around this subject. You trump everyone, even the integrity of this feature. . . . Remember, whilst this card may be good for the purposes of the game, it's bad for the purposes of society at large."

Although, having accused Mr. Thompson of being humorless, I must admit that the tags he used for his piece did make me laugh: "Tags: Islam, *New Humanist*, politically correct atheist cowards." Now that is funny.

Because of the interest, the fuss, the fantastic reception, and multiple requests for more cards, I ended up creating a second set to accompany the originals, the intention being to ultimately produce a full pack of playable cards, which would be distributed by *New Humanist.*

The second set was previewed in the March/April issue and sparked yet more requests for further sets. At the time of writing, I think we'll be leaving it at twenty-four cards, but who knows? With more than 4,000 differing belief systems filling up the world, maybe we'll produce more eventually.

Designing the Atheist Bus Campaign

GRAHAM NUNN

Had it been necessary to apply for the position of Atheist Bus Campaign designer, I'd never have got the job. My CV, in a brief review en route to the wastepaper basket, would have shown no previous experience of design work apart from an apologetic mention under "hobbies," and even that would have been a vain attempt to pad the whole thing out a bit. I'm the sort of person who needs to type his CV in an oversize font just to ensure the need for a staple.

I was therefore rather fortunate that it all fell into my lap. The organizer, Ariane Sherine, in a couple of comment articles for the *Guardian*, had already garnered some interest in the idea of running an alternative to the religious adverts that were appearing on London buses at the time, some of which indirectly threatened all non-Christians with eternal torment in hell. As if public transport wasn't bad enough. Her proposal was to respond with the cheery slogan "There's probably no God. Now stop worrying and enjoy your life." It wasn't an altogether serious notion at first, but it became so as more people registered their support and asked how they could make it happen.

It was a good question. Why had nobody done it before? Perhaps they'd considered the idea of getting a large group of freethinking non-believers to work together and decided that their efforts would be better focused on tackling simpler tasks, like giving pigs the power of flight. In any case, Ariane persisted with the proposal and, knowing that I was reasonably competent with a computer, asked me if I could lend the fledgling campaign some visual impetus by creating a mock-up of how the advert might look on a bendy bus. These serpentine vehicles were cheaper to advertise on than normal buses and were therefore a realistic target financially, despite being universally loathed by pedestrians and road users alike.

My brief was simple enough—to produce a clear, stark representation of the slogan using Ariane's choice of colors. I duly selected a rugged-looking font, gave her a few layout options and, once the ad's appearance had been settled upon, applied it to a bus photograph. And that was that. I was pleased that this image would be used for promotional purposes but didn't anticipate any further involvement. Even if by some non-divine miracle these bus ads ever made it onto the roads, Ariane would obviously get a proper designer in for the serious stuff. With no experience of such matters, I naively assumed that you can't produce final advertising artwork on a modest PC in your bedroom. Surely there were secret requirements that take such powers away from hapless amateurs like me? It was probably for our own good.

Anyhow, nothing was going to happen unless enough money was raised on the official donation site when it went live. This was where the idea could easily fall flat on its face, as members of the public had to be persuaded to part with hard cash. Despite the many positive comments that had been posted online, it wasn't easy to gauge whether enough people were prepared to open their wallets, especially as the response to an earlier pledge scheme had been lukewarm. Seizing on this early event, the *Daily Telegraph* had run the headline

"Atheists Fail to Cough Up for London Bus Ad" and pronounced the idea dead with little chance of resurrection. There was soon reason for optimism, though, as the British Humanist Association had offered to administer donations to the campaign, and Richard Dawkins had also declared his support by offering to match every donation up to the target of £5,500—which meant, effectively, that only half that sum was required in total from everybody else.

Even so, I tried not to get my hopes up as I logged on for the first time that morning. According to the media, the credit crunch was stomping all over the financial markets like a video game baddie and bankers were fleeing for their bonus lives. We were all feeling the pinch. Surely a bus advert would be seen as an unnecessary extravagance while the nation collectively tightened its belt. The answer was emphatic. On *Blue Peter* they used to make elaborate "totalizers" for their appeals, which crept up slowly week by week in true "will they, won't they?" goal-reaching drama. Had their props team been charged with building one for the bus campaign, they would no doubt have been a bit miffed to see it become redundant before they'd finished packing up their tools.

The suddenly modest-looking target was met with almost embarrassing ease, but far from tailing off, the rate of contributions kept increasing. Administrators for the donation site reported unprecedented activity as people from all over the world gave generously, and in many cases repeatedly, leaving supportive messages as they did so. The "herding cats" analogy that is sometimes applied to atheists had been made to look rather foolish, along with any lingering doubts, as the total soared to over £100,000 in just four days. Eat that, *Daily Telegraph*! It was now clear that not only were the buses going to happen, they were going to happen on a much grander scale than anyone had dared to anticipate.

The upshot of this was that instead of having 30 buses in central London, there would now be 800 all over the UK. Bus spotting—long the preserve of bespectacled cagoule-wearers wielding spiral-bound notepads—was now a genuinely tantalizing prospect for atheists across the country. It was fitting that many more people would get to see them, given that donations had come in from far and wide.

In spite of my doubts, I remained the official designer. I believe that, due to Ariane's misplaced cheerleading on my behalf, the British Humanist Association were under the impression that, far from being a nerdy enthusiast with some expensive software, I did this kind of thing for a living. It was a fair assumption on their part but laughably inaccurate all the same. I wasn't about to correct them, though. After all, the only difference between me and the professionals was that they'd gone to the terrible inconvenience of attending university and gaining a degree in the subject. Who's got time for all that nonsense?

I was gaining in confidence and started to feel less out of my depth. I don't wish to denigrate the work of professional designers, of course—you've only got to look at some of the adverts in local services directories to see the embarrassing results of the "why pay someone when we can do it ourselves" approach—it's just that it wasn't the hardest of jobs to tackle. Align some text, change the colors, try not to spell *God* incorrectly . . . okay, so there was slightly more to it than that, but nothing worth boring you with. It was still reasonably straightforward, and by now I was in receipt of the specifications from the advertising company, which were surprisingly straightforward and revealed that there weren't any secret requirements after all. Thinking about it now, it would have been madness to pay a proper company an exorbitant fee to re-create what I'd already done.

As interesting as the bus advert, though, from my point of view, was the decision to spend some of the extra funds on adverts for the London Underground network. There would be four different versions, each featuring a quote from a notable atheist, which meant that more design work was required. And so I set to work again, amused by the idea that people might ponder these adverts while their faces were buried in the armpit of a fellow passenger during rush hour. Well, any distraction had to be a good one, right?

On the day of the official campaign launch, I armed myself with a cheap camcorder and hopped on the train to London. I was wearing my badge and T-shirt, which had also fallen under the design remit, and felt pleasingly out of place amid the suited commuters around me. The temperature would barely nudge above freezing all day, but a heated marquee awaited the gathering media presence, which included not only national press but correspondents from all over the world. I was meant to bring a large banner with me to hang above the stage, but unfortunately it never turned up, and the banner company claimed the courier had "lost" it. I might have believed them, but this happened not just once but twice—so I reckon they'd taken some kind of anti-atheist umbrage to it and tossed it in the Thames (although it's more likely that the banner company had refused to print it in the first place and lied). This was a shame, not least because Ariane completely forgot that it wasn't hanging behind her and mistakenly alluded to it in her speech.

The launch itself was a great success. On the stage we had set up easels with enlarged versions of the four different tube adverts, which were unveiled individually by the guest speakers. The mood was positive, and the complementary T-shirts were more popular than the complementary snacks. Afterward I learned that some of the press had been asking which design company was responsible for the artwork. Presumably they were going to mention it in their articles, but on learning that it was some tin-pot chancer working in his bedroom, they obviously thought better of it.

Before returning to Suffolk I made a point of stopping off at Oxford Street. The post-Christmas sales were in full flow, but it wasn't bargain housewares that had lured me there. Opposite the Bond Street tube station were two animated screens displaying our slogan on a thirty-second rotation with the other ads. I'd knocked up the animation myself after hastily learning the basics and, to my eyes at least, it looked really good. It was the day after Twelfth Night, and it seemed like a fitting replacement for all the Christmas light displays. No one was paying much attention to the screens, apart from me with my camcorder as I failed not to look like a weird tourist, so they probably gave me more satisfaction than anything else. These animations were later used in news reports both here and abroad as passers-by were stopped in front of them and asked for their opinions. A few shocked old ladies are always good value for a vox pop. And that, of course, was the point—not to shock old ladies, but to get people talking. Shocking old ladies isn't big or clever.

The media coverage following the launch was extensive. The unusual nature of the campaign proved irresistible, and the story was featured on television and radio stations all over the world. While this generated extensive debate, there were only three notable adverse reactions. First, a marginal group called Christian Voice (who, somewhat ironically, I had never heard of up until this point) tried to get our adverts banned but succeeded only in exposing their general intolerance and giving us more publicity. Their leader, Stephen Green, cut a desperate figure as he spluttered his disapproval from a soapbox around which no one appeared to be gathered.

The second vocal objector was Ron Heather, a Christian bus driver from Southampton who had refused to do his job after seeing one of our ads on the side of his trusty steed one morning. If Christians were looking for a heroic figure to defend their faith against the terrifying onslaught of—*scream!*—an alternative viewpoint, they couldn't have done much worse than the limp-as-lettuce Mr. Heather. His defiant gesture was to wander home and call BBC Radio Solent. He enjoyed his fifteen minutes in the spotlight but didn't do much to further his cause. If I were being childish, I could point out that he is one pen stroke away from being Mr. Heathen, but then with my surname I'm not in the best position to do so.

The third and most interesting response—for me at least—came from a political group called the Christian Party. Their leader, George Hargreaves, was someone I remembered from the Channel 4 series *Make Me a Christian* in which he had tried to imbue various unlikely candidates with a more godly outlook on life while confiscating "unhelpful" possessions such as sex toys and books on witchcraft. His success had been rather limited, to say the least, but perhaps he had agreed to the program with a wincing nod to his own past; prior to his foray into ministry he had been responsible for writing and producing pop records, among which was Sinitta's gay disco anthem "So Macho." Now, I would wholeheartedly agree that this act alone required a great deal of forgiveness, but becoming a reverend was perhaps one apology too far.

Anyway, it transpired that he didn't like our adverts very much and had taken it upon himself to launch a riposte. Seemingly missing the point, as many had done, that our adverts had been a response themselves, his considered effort was to run his own bus adverts that copied ours almost entirely. The colors, the font, the layout . . . the only difference, in fact, was the wording, which now read: "There definitely is a God. So join the Christian Party and enjoy your life." My initial reaction was one of amused confusion. Was this a brilliantly conceived appropriation or the workings of the least imaginative man in Britain? Was this copycat campaign just a case of So Match-o?

The answer could be found in a statement from the man himself on the party's website. After acknowledging that his tolerance had been tested by our slogan, he quoted from Proverbs 26:5: "Answer a fool according to his folly, lest he be wise in his own eyes." In other words, decry a rational and positive statement with one that can only be digested with a heaped spoonful of blind faith. Folly indeed. And what about "Thou shalt not steal"? Perhaps I'd missed an exemption footnote. At least ripping off our advert meant that he didn't have to design one for himself, which you have to admit was clever if somewhat lazy. All he needed was a mate with the right software. Maybe Sinitta had done it for old times' sake.

A few people asked me if I was going to take action for copyright infringement, but I think that would have been rather petty. At a glance, the Christian Party adverts looked like ours anyway, so if people were recalling the original slogan when they saw them, it was hardly doing us any harm. Some things just ridicule themselves, and so we did the decent thing and turned the other cheek. There's a tip for you, George. Now give the nice ladies their dildos back and stop being such a spoilsport.

Atheist buses weren't unique to the UK. Taking Ariane's lead, other countries ran similar campaigns, including the United States, Canada, and Spain. Attempts to run ads in Italy and Australia were shelved following resistance from advertising authorities, but efforts persist. It was a pertinent reminder that the freedom of speech we take for granted in the UK isn't as common as we might imagine. The ads in the United States carried their own slogan ("Why believe in a god? Just be good for goodness' sake"), but those behind the campaign in Canada asked if they could use the same slogan as we had. I happily sent them the artwork and was delighted to see photographs of it paraded on buses in Calgary and Toronto.

On reflection, everything connected to the campaign artwork went according to plan, and I think I got away with it. As far as I know, nobody laughed and said that the design was the work of an incompetent buffoon who should never be allowed near a computer again. That's good enough for me. I'm well aware that any number of people could have done the same job to an equal or higher standard, but no one was unkind enough to point that out. No one said much at all really, although a work colleague of mine recently told me how one of the official car stickers I gave him had scared away some Jehovah's Witnesses from his front door. If that's not a positive result, what is?

The Little Atoms *Radio Show*

Neil Denny

It is an oppressively hot summer evening, but we are immune from the heat outside, sequestered in a cool, air-conditioned basement below the School of Life in Bloomsbury, the philosophical heart of London. *Little Atoms*, the radio show I produce and co-present with Padraig Reidy, has been running for nearly four years now, and we have decided to embark upon a bold and possibly foolhardy experiment. The show is normally broadcast live from the Resonance FM studios, with just a guest, an engineer, and ourselves in the room—but for the purposes of recording tonight's show, we have hired a venue and made tickets available, and the writer Jon Ronson has kindly accepted the invitation to be our guest and guinea pig for the evening. We are about to share with an audience the arcane magic that goes into making our radio show. That magic basically consists of two people with hastily scribbled notes talking to a third person, and me periodically pressing stop and start on my MP3 recorder. Still, in the comfort of the studio it is easy to pretend that nobody is actually listening—but there is no such escape today. They are here to listen to us discussing Jon's latest film, *How to Find God*, which charts the journey of a number of agnostics toward "spiritual salvation" through the medium of the Alpha Course. The tickets for the interview were snapped up within twenty-four hours of being made available, and we have a waiting list for cancellations. An air of expectancy fills the room.

The seeds of the idea that became the *Little Atoms* radio show were sown in a pub garden by London Bridge on the afternoon of July 7, 2005, while I and a skeptic named Richard Sanderson waited for the trains to start running again. Richard and I had been introduced by a mutual friend, and we had been going along to Skeptics in the Pub together on a regular basis. Both lifelong atheists, we had often discussed working together on a radio show about science and rationalism. That afternoon we discussed how, while London came to terms with the fact that it was once again the target of terrorist bombs, nominally progressive commentators had already begun to pontificate online that the blame for these lay with the government. The idea that this might have been the personal responsibility of individual religious extremists was anathema to them.

This response served as the final catalyst. We decided that very afternoon to create a show that would actively promote the intellectual legacy of the Enlightenment and would defend freedom of expression, free enquiry, empirical rationalism, skepticism, the scientific method, secular humanism, and liberal democracy. At the same time the show would cast a critical eye over the *bien-pensants* who automatically attribute all the world's ills to the actions of "the West," over the purveyors of religious dogma, superstition, and magic potions, over those nostalgic for totalitarianism, and over the adherents of deranged conspiracy theories.

We pitched the idea to Resonance FM, suggesting ourselves as balance to *The Headroom*, a syndicated show presented by a UFOlogist, which was at the time their most popular show—and so, in September 2005, *Little Atoms* was born. In August 2006 Padraig joined the show as a presenter, having previously been a guest. Over the past four years we have been joined by an array of interviewees, including Christopher Hitchens, Jonathan Meades, Noam Chomsky, Ann Druyan, Alain de Botton, Adam Curtis, Julie Burchill, Marcus du Sautoy, Francis Wheen, and Richard Holloway, as well as quite a few of the contributors to this book.

So far, the night has been a great success. Much wine has been consumed, Jon is in great form as usual, our questions have been reasonably erudite, and the audience is laughing along with the interview. We take a brief pause so that I can think of how to phrase a link to the next part of the interview. Jon surveys the room, thanks everyone for coming, and then declares, "I'm really excited to see this cornucopia of skeptic celebrities in the room . . . it's like a kind of skeptical heaven!"

We should probably take a few moments at this point to discuss the tenets of skepticism. What do people who self-identify as skeptics actually believe? Aren't they just, well, skeptical?

It clearly isn't practical for us to live our lives questioning everything. As somebody wise once said, "If you're too open-minded, your brain will fall out." There are distinct parameters and techniques that can be learned to help us lead a more skeptical life. Let's look at a few examples of things that a skeptic should (and shouldn't) be skeptical about.

A skeptic is skeptical about paranormal phenomena: the existence of a God or gods and the afterlife; creation and reincarnation; ghosts and spooky visitations; psychic phenomena such as extrasensory perception, remote viewing, and fortune-telling; the supposed ability of a medium to communicate with your dead grandmother or hamster; or the ability of a "psychic soldier" to stare a goat to death.

A skeptic is open-minded about the possibility of life on other planets but is also quite confident that midwesterners and their cattle are safe from the invasive medical practices of Venusians. Likewise, a skeptic is not convinced by shaky camera footage of Bigfoot or the Loch Ness Monster.

A skeptic worries about the rise of alternative medicine and its use within the National Health Service, the efficacy of magnet therapy, reflexology or crystal healing above and beyond the placebo effect, homeopathic malaria pills, chiropractic cures for childhood ailments, or the use of vitamin C as the sole treatment for AIDS.

Nonetheless, a skeptic should also be skeptical about the sharp practices employed by big business, the drug, tobacco, and oil industries; about the fake "controversies" invented by advertising, PR, and government spin; and about the mass media and its selection bias and love of fake "experts" and food gurus.

But a true skeptic should also have a keen nose for bogus skepticism, which often manifests itself in the form of conspiracy theories.

Why oppose conspiracy theories? Isn't it the province of the skeptic to be skeptical about man landing on the moon? Or of HIV being the cause of AIDS? Or of MMR being a safe vaccine? What about the crackpot idea that "nineteen men living in a cave in Afghanistan" could have had the nous to bring down the World Trade Center? Or that life on earth could have evolved through natural selection?

The key variable to note here is evidence. True skeptics have at their disposal an amazing array of tools that have been honed by great thinkers across many civilizations down through the ages—tools like the scientific method, controlled experiments, and double-blind testing. The skeptic learns, in the words of Carl Sagan, "to construct, and to understand, a reasoned argument and—especially important—to recognize a fallacious or fraudulent argument . . . [and to] recognize the most common and perilous fallacies of logic and rhetoric."

(Much more of this can be seen in the chapter "The Fine Art of Baloney Detection," from Sagan's magisterial book *The Demon-Haunted World*—if you don't happen to own it, you should put down this book right now and go and buy it!)

The other key point to note is that, unlike with the dogmatic belief system of the religious zealot or the conspiracy theorist, if at some point in the future compelling scientific evidence comes along, say, for the existence of God or chakras or Nessie, then skeptics will be obliged to change their mind.

After the interview there is a lively question-and-answer session with the audience, and then a group of us retire to a local pub. I ponder what Jon had said earlier about the skeptical luminaries in the room. It occurs to me that while the observation he made was correct, a large number of these people are also my close friends. We regularly socialize together and attend each other's events and meetings. Is this really suggestive of a growing interest in skeptical thinking among the wider public? The inner skeptic in me wonders if we could be just a small group of self-aggrandizing obsessives who constantly reinforce our own belief system to each other. Is there actually any evidence that this stuff matters to people?

To answer that question: the past couple of years have seen a number of significant initiatives, events, and happenings in the skeptical world.

Most people reading this book will be aware of the spate of best-selling books on atheism that have hit the shelves in recent years. Richard Dawkins and Christopher Hitchens in particular, both well-respected writers in their own fields, have written hard-hitting books on God and religion that have massively outsold most of their previous work.

To top off a year that saw the Atheist Bus Campaign, many successful Skeptics in the Pub meetings, and the Nine Lessons and Carols for Godless People concerts, now to become an annual Christmas fixture, October 2009 saw the first London iteration of the Amazing Meeting, the yearly conference of skeptical thinking run under the umbrella of the James Randi Educational Foundation. This two-day conference, featuring talks given by some of the world's leading rationalists, was attended by around 500 people, and had sold out within twenty-four hours of tickets going on sale.

One setback for skeptics this year has been the ruling by Lord Justice Eady in the libel case pitting the British Chiropractic Association against Simon Singh. Eady held that by using the word *bogus* in a *Guardian* comment article to describe the BCA's promotion of chiropractic for the treatment of a number of childhood ailments, Singh was stating as a matter of fact that the BCA were being consciously dishonest. As this was not what Simon had been arguing, it made his position difficult to defend. At the time of this writing, Simon had made the decision to appeal.

However, even this disappointing ruling has seen a positive and optimistic outcome: the science advocacy group Sense About Science launching a campaign called "Keep Libel Laws Out of Science." An online petition received 15,000 signatures within a few days, including a large number of prominent people from the arts and media as well as scientists and skeptics. At the same time, a loose coalition of science bloggers began to closely investigate the claims of their local chiropractors, which saw a large number reported to local trading standards departments, an action that caused a rival chiropractic association to claim without irony that they were victims of a "witch-hunt."

The examples above seem to me to be ample evidence of two things: first, that there is a huge appetite for atheism, skepticism, and rational thinking out there, and second, that we organizers of skeptical events need to start booking bigger venues! Indeed, the very fact that you have bought this book and are reading these words suggests as much. Of course, as a skeptic, I'm willing to accept that I could be wrong, and that perhaps the only people who are reading my words are the same people who sat in that basement back in the summer. If this is indeed the case, then happy Christmas, friends! I'll see you all in the New Year for a pint.

James Randi:

The Real Santa Claus

Sid Rodrigues

I have very, very religious parents. Our family originates from the west coast of India, where the surname Rodrigues gives you a clue as to the Portuguese colonization that took place in the sixteenth century. All my ancestors were Hindus, and the Hindu caste system in India is very rigid—you're born into it and can never ascend or descend within the system or marry outside of it.

At the time, there were only two ways to get out of the caste system: either you could pretend that you weren't from that caste, or you could convert to Christianity. The first option didn't work particularly well, because you'd have to try to hide away from your past, and it could always come back to haunt you—particularly if you wanted to get married and had to prove you weren't illegitimate.

So in the sixteenth century a lot of Hindus on the west coast, including my family, converted to Catholicism (hence my real name being Simon-Peter, not Sid). I went to a very Catholic school where we were called by our surname, but at the age of five, most couldn't pronounce "Rodrigues," and unfortunately there were a few others in the class called Simon or Peter! (Not to be confused with Peter Simon, the eighties kids' television presenter of such compelling Saturday morning shows as *Double Dare*.) So in the end, my first name was shortened to Sid, and I've been called that ever since.

My Catholic primary school was full of teachers who were great fun, and the good thing about going to a Catholic school was that we always got lots of time off for a quick pray in the mornings, and went to mass on what seemed like random occasions, which were actually feast days. As I hadn't thought about it much at the time, I went along with all the religious procedures and prayed a lot.

Then, when I was about twelve, I developed a keen interest in magic and conjuring, and was also a bit of a science nerd. Predictably, I was transfixed by a program on television called *Psychic Investigator*, which featured a magician called James Randi. I never missed an episode, and loved to watch the Amazing Randi apply science to all these "psychics," like Uri Geller, who purported to have magical powers. I hadn't thought you could test psychics, but James showed you exactly how the scientific testing worked, and so my interest in skepticism began.

At the end of the series, because I'd enjoyed the show so much, I took the initiative. The show's producers had brought out a book, and I rushed out and bought it with my pocket money. In the back of the book there was information on a challenge offering a $10,000 prize for anyone who could demonstrate paranormal powers under scientific conditions, so I wrote to the address printed there on the off chance that James Randi might write back. That letter started off a correspondence across the Atlantic that has lasted for nineteen years.

A few years after we started writing, we got onto the subject of God. James knew about my Christian upbringing straightaway, because my name lent a clue that I was from a Christian family, and asked, "Are you Catholic?" Even though he'd clearly worked it out from my name, at the time I thought he'd managed to read my mind. I explained that yes, I was a Catholic, because I'd been brought up that way. He never questioned this, but occasionally he would point to different bits of the Bible and say, "Does that make sense to you?" I know now that he was teaching me to think for myself.

Once James had got the ball rolling, I started to do my own research, and slowly stopped believing. However, because of my family, I can still fully understand that religious belief is extremely comforting. It also involves a significant emotional and financial investment over the years. The older you are, the more of your life you've invested as a believer, and if you decide to give it all up a long way into your life, it probably makes you feel a bit daft. Because I was so young, I had none of this baggage to hinder me.

A lot of people I know who are atheists have parents who weren't really that bothered about religion, but if your parents are devout, it's much harder to make that leap. I don't know what mine made of these letters from James Randi—they either thought it was a bit odd or decided it was quite nice that I had a new friend. I really looked forward to his letters, which would always turn up on different types of novelty paper. One would look to be printed on a crumpled sheet of paper; another would be printed on the back of the final draft version of his latest book's glossy dust jacket. Every time I received a letter, it would be like Christmas all over again.

I was pen pals with James throughout my teens, but I'd entered university before I met him for the first time. I received an e-mail from him saying: "What are you doing at the weekend? I'm in Bristol, and am heading up to Oxford to meet a friend and do a talk there. I know it's quite close to London, so is there any chance you could meet me in Oxford? We'll sort you out a hotel there, or you could stay at Richard's place—I'll give you his number." I was incredibly excited. I had no idea who the "Richard" he was referring to was, but duly booked my ticket and arrived in Oxford. I needed to get directions, so I called the number he'd given me and talked to the man on the other end. It was only once I had hung up that I realized I had been speaking with none other than Richard Dawkins!

I went to a pub to have some lunch and a beer before setting off on the second leg of my journey. As I was sitting at the table, it dawned on me that today I was going to meet two of my heroes. As I left and started the walk to Richard's house, I started getting really nervous. I knocked on the door, my stomach churning, and his personal assistant answered and asked, "Can I help you?" I was a gibbering wreck and couldn't even speak. "You must be Sid, then," she said. "Come on in—they're both here!"

It was both bizarre and wonderful, as James Randi and Richard were both very welcoming. James was nearly seventy by then, and his honesty and open-mindedness made a great impression on me, since he always said exactly what he thought. I sat and listened to him, Richard, and Richard's wife, Lalla, talk about science and skepticism, but I was so in awe I barely said a word all weekend.

Their talk was the most memorable skeptical event I'd ever been to, and it was one of the best weekends of my life, spent over pasta and science. I remember getting progressively drunk on wine, but, thankfully, I don't remember saying anything silly. I arrived home startled by how scientists and skeptics could encapsulate everything I'd been thinking in such a clear and succinct manner. I had been completely won over.

A short while later, while I was still at university in London, I stumbled across an event called Skeptics in the Pub—which just happened to take place just over the bridge from where I was studying. If I wasn't such a skeptic, I'd say it was serendipitous. Skeptics in the Pub was run by an Australian named Scott Campbell, who had come up with the idea of organizing monthly meetings for people interested in skepticism, and it soon became one of my unmissable events. A few years after I began attending, Scott handed the night over to a guy called Nick Puller, who ran it for five years before moving to Hungary, at which point the reins were handed over to me. I've now been organizing and running it for the past two years, in which time it's grown from a forty-to-sixty-person meet-up to what *Time Out* has certified as "probably the world's biggest pub meeting."

The first thing I did when I took over was to move it to the biggest function room I could find: the Penderel's Oak in Holborn. It holds up to 300 people and is often packed. I run the night once a month, but sometimes we'll put on a special extra night at short notice if a notable skeptic is in town and likes the idea, as we can spread the word quickly through Facebook and Twitter.

It doesn't always go smoothly, and there's even rumored to be a "curse of Skeptics in the Pub"! At one of the first events I went to in Waterloo, ten minutes into the speaker's talk, there was an electrical short because of a road digger outside, and the whole pub was plunged into darkness. Someone smelled burning, and we realized that a small flame was slowly illuminating the back of the bar where a fridge had caught alight, so the whole pub had to be evacuated. A couple of people arrived late and tried to get into the pub, and the landlord said, "I'm sorry, but the pub's burning down—you'll have to come back next week!"

Then, when we moved to Holborn, a speaker was halfway through a talk when a waitress ran downstairs and said, "I've got an announcement to make—does anyone own a car with the following registration number?" Nobody did, so she scuttled away, only to come back five minutes later and say, "Sorry, we've got to evacuate the whole building, because there's a bomb alert and the police are cordoning off the whole road!" So, ironically, they led us outside, straight past the dodgy suspected bomb car—if they'd left us in the basement, we'd have been a lot safer!

Until around two years ago, London was one of the only venues for Skeptics in the Pub, but then I gently persuaded my long-distance partner, Rebecca, to open one up in Boston, Massachusetts—and now they're all over the place. There are nearly thirty regular events in the UK, the United States, and Australia, furthering rational thinking across the world.

Meeting Rebecca was one of the best things about becoming a skeptic. James Randi has a conference in the United States called The Amaz!ng Meeting (TAM), and I was lucky enough to be invited to the first one, where a lot of brilliant skeptics from around the world shared stories and had a laugh together. A few years later, TAM 4 took place in Las Vegas, and on the first day I met my future wife. We went out in the sun and made all the delegates' name tags and had a chat, and because the weather was so hot she jumped fully clothed into the swimming pool—I immediately fell in love. Because we met at TAM 4 in Vegas, we decided it would be fitting to get married at TAM 7, which also took place in Vegas, and so in July 2009 we tied the knot in front of all of our fellow skeptics, like one big skeptical family. Our wedding video is on the intertubes, and Jon Ronson comically called us the John and Yoko of skepticism. (He'd originally called us the Brangelina of skepticism, so it was a vast improvement.)

Letters have given way to e-mail and Skype, but I will always be grateful to my bearded skeptical Santa Claus, the Amazing Randi—because without him, none of this would ever have happened.

SID'S TOP TIPS FOR GETTING INVOLVED IN SKEPTICISM

It's predictable, but I'd say that if you can, try to get to Skeptics in the Pub. In the UK, there are monthly events in London, Oxford, Leicester, Leeds, Liverpool, Birmingham, and Edinburgh, so you'll never have to feel alone in your skepticism again.

If you can't get to those, there's always the James Randi Educational Foundation message boards and the UK Skeptics message boards, and there's also a great forum at richarddawkins.net. Richard Dawkins even joins in the discussions on his forums sometimes.

Read books on skepticism, such as Ben Goldacre's *Bad Science*, Richard Wiseman's *Quirkology*, Phil Plait's *Death from the Skies*—and, of course, anything by James Randi, who now contributes to an online blog called Swift, which is published at www.randi.org.

Join the British Humanist Association or the National Secular Society, as they have lots of local area groups you can join. Humanism and secularism are different from skepticism, but both types of group will be made up of rational people who believe in the natural world rather than the supernatural. You can also look on meetup.com for nearby groups in your location.

A Day in the Life of a Godless Magazine

CASPAR MELVILLE AND PAUL SIMS

New Humanist MAGAZINE

Flecks of snow, each one resplendent in its absolute uniqueness, swirled around his head, and the charred aroma of chestnuts roasting on an open fire filled his nose as the young news editor made his way through the now almost deserted street of the city to his cramped desk in the Blasphemy Lab, high up in the draughty garret of Godless Towers, for a century the headquarters of English heresy.

It was Christmas Eve, and he was late. Late for the crisis editorial meeting called by his splenetic but brilliant boss, Ebenezer Jazzfunk, the editor of *Unbelief Bi-monthly*, the market leader among humanist and rationalist magazines in this part of the country, with but one modest mission: to dismantle several millennia of monotheism. Ebenezer would not be pleased.

Squeezing past the overflowing boxes of past issues and inexplicably unsold copies of *Is a Fideistic Theology Irrefutable?*, he tried in vain to slip unnoticed into the meeting room, where the emergency session had already commenced, but an intemperate advancing of the door knocked loose an errant box of copies of *Rationalism and Population Explosion: Was Swift on to Something?*, which hit Jazzfunk square on the head, alerting him in no uncertain terms to the belated arrival of his assistant.

"So, you are here!" the editor boomed, somewhat redundantly. "Well, catch up, boy." The meeting, he explained, constituted an attempt to avert the single biggest crisis in humanist publishing since the Battle of Conway Hall. The problem? Well, it concerned the Christmas issue, more properly (though perhaps less pithily) known in these part as the "late December annual holiday celebration issue." Weeks of assiduous labor, the editor explained, had dredged up an issue of quite brilliant erudition. There was the compelling memoir "My Life as an Ardent Atheist but Also Extremely Moral Person," by Cynthia Puffer; the debut of a new rationalist advice column, "Just Bloody Well Grow Up and Stop Moaning"; and, lest we forget, this was the issue that would see a special children's supplement titled "Why Mom and Dad Lie: Ten Reasons Father Christmas Definitely Does Not Exist."

"But," roared the editor, as the skeleton of a pigeon, knocked loose by the snow, fell into the fire grate, "we have still not proven the nonexistence of God. We lack that killer lead story, that hook-laden, pithy tour de force that will finally, this December 25, succeed in wrenching the Christ from Christmas. And you," he said, turning to the news editor, "are the man for it. As you know, for very rational historical reasons I cannot now recall, we keep the front page open until last thing on Christmas Eve for the lead story. You are to find that story. You are to steal Christmas. And see that you do or you will find yourself seconded to the Humanist, Rationalist, Agnostic, and Ethical Kindred Liaison Committee come January."

And with that (and ten minutes spent stuffing papers into his bag, struggling into his tweed overcoat, and trying but failing to conceal from his trembling but nonetheless scornful assistant the fact that his pockets were stuffed with Christmas presents for his grandchildren, all neatly wrapped in paper bearing small but still distinguishable pictures of wise men, angels, and a cowshed with a comet overhead) the editor was gone.

Christ Almighty! the news editor thought. *I'd better get on with it.* And with that he immediately went to work, employing that trusty time-honored journalistic research tool, Twitter. It wasn't long before his plaintive cry was popping up on Tweetdecks across the metropolis: "I've only got eight hours to prove that God doesn't exist—any ideas?"

Almost instantly a message was fired back: "Walking down the high street, for my usual Twinings lapsang souchong and a madeleine. Snow is ethereal, isn't it?" It was clear that Stephen Fry was going to be no help.

His mind wandered back over the eventful yet strangely tedious year he had spent at *Unbelief Bi-monthly* after graduating with a double second from Bejesus College, Cambridge.

Time was short. *Think! Think!* He saw the telephone. For some reason it reminded him of a conversation he had had, not three months previously, on a telephone. He'd been in charge of answering the office phone ever since the receptionist had not been hired yet. It had rung insistently before the news editor obliged and answered.

"*Unbelief Bi-monthly.*"

"I've received a copy of your magazine," said the lady on the other end. "And I demand to be removed from your mailing list at once."

"Well . . . er . . . that's no problem at all," said the news editor. "Can I take your name . . . ?"

"Well, it wasn't addressed to me. I object to receiving something through my door that insults God in this way."

"Whom was it addressed to?"

"The man who lived here before me. He passed away, and I dread to think what has happened to his soul. How dare you send something like this to me!"

"But it wasn't intended for you and—"

"I don't know how you live with yourself."

"I'll ensure you're removed from the mailing list. I'm sorry you received it."

"It's not me you should be apologizing to!" She slammed the phone down.

Remembering that call, he knew the telephone would be no use. Glancing up at the office clock, which though fairly accurate in terms of time had all the numbers in a random order because it had been bought on the cheap from the shop nearby run by a blind watchmaker, the news editor realized time was running out.

There was nothing for it. He would have to risk everything and delve into the green ink file. This file, more of a folder really, or ring binder, or plastic wallet, had been treated by a succession of editors mainly as a scribble pad and a place to put cups of tea, and was basically ignored when it wasn't being openly denounced. This was where generations of godless editors kept their holy grails: the readers' letters.

The news editor felt trepidation as he ferreted under a pile of Norwegian Humanist brochures and the journals produced by the formidable FreeSecs Union, formed by a recent merger of the National Freethinkers and Ethical Secularists. The green ink file was sacrosanct, but that alone was no reason to befoul it. Yet he had no choice. If the final proof of God's nonexistence existed, it would exist here, somewhere among the unpaid bills and cries of outrage at that article claiming that wearing bright red lipstick was a profoundly humanist act.

The news editor spotted some familiar handwriting. The editors had long ignored the weekly postcards whose author signed not with a name but a symbol and which proclaimed, "Blood and urine tests prove we come from apes and that all holy texts are lies." Given his sacred quest, the news editor read the headline with a new urgency. He felt an almost otherworldly sense of destiny.

"I have just today," said the card, "asked the Home Secretary to arrest the Archbishop of Canterbury and the Queen for deception. In my letter to the Home Secretary I explained how blood and urine tests show that we come from apes." Pay dirt.

But his elation turned to dust and ashes when a call to Scotland Yard revealed that both the Queen and the Archbishop remained at large.

Then his attention was caught by a letter headed "Suggestions for improving *Unbelief Bi-monthly*." "Dear Editor," the letter began. "*Unbelief Bi-monthly* is not sufficiently sexy or lighthearted, when one considers newspapers such as the *Daily Sport*, the *Sun*, and the *Daily Star*, which all contain pictures of pretty girls. I hope *Unbelief Bi-monthly* would become more sexy/humorous, in line with society in general." Attached to the letter were clippings from various publications, which did indeed contain pictures of pretty girls. Could this be the answer the news editor was looking for?

After briefly considering a humanist Page Three, instinct told the news editor to move on, and he dug into the mailbag once more. He skipped past the one titled "Four Hezbollah Anagrams" (Hello H Baz, L H Haze Lob, Ha Boz Hell, and Blah Z Hole) and another requesting a written reply proving that God didn't exist ("I'm sorry to have written a letter, but sadly I do not have an e-mail machine") and alighted on a letter whose author claimed to have hit upon proof of creationism—perhaps this would be the biggest scoop of all. Imagine if the news editor, in his quest to prove the nonexistence of God, instead unearthed proof of God! "Most people think that fossils are proof of evolution, but this is not the case," explained the author. "Fossils and dinosaur bones are a side effect of creation. When God created mankind and other creatures, the fossils automatically appeared because the created world must work logically. Dinosaurs never existed but dinosaur fossils do."

The news editor was beginning to despair. It was 3:00 p.m. on Christmas Eve, and he was no closer to establishing how this godless magazine was going to prove God's nonexistence. Taking one final dip into the mailbag, he pulled out a letter from a prospective contributor to the magazine, offering a piece of fiction for publication. *Unbelief Bi-monthly* rarely published fiction, but he read on in hope. The brief synopsis carried the title "The Turtles and the Gulf Crisis" and described itself as "a story about the Teenage Mutant Ninja Turtles and the Persian Gulf crisis." The author had taken the trouble to explain the leap of imagination that had led him to this tale: "The reason I have put the Turtles and the Gulf crisis together is because they seem to go together. They are contemporaneous and seem to share the same mentality. The story is told in a serious tone, not tongue in cheek. I think it is more effective that way."

It was no use.

The final editorial meeting was fast approaching, a meeting that Ebenezer Jazzfunk had proudly declared, on his recent Sunday morning appearance on the BBC's flagship faith show *Debates Between People Who Will Never Agree*, would be "the final editorial meeting of the theistic age." The news editor scuttled into the meeting room and prepared for Jazzfunk's final judgment.

He found the editor in resigned mood. "Don't tell me," he said. "You didn't find it. Don't worry, lad, it's the same every year. We'll just have to go with the same front page we ran last year—'Jolly Festiwintervus to Everyone.' Now, who's for eggnog?"

Note: While the story itself is fictional, all correspondence listed was genuinely received by *New Humanist.*

Front Line in the War on Christmas

Andrew Shaffer

"Merry Christmas or happy holidays? Which strategy should retailers use to cash in? Here for a fair and balanced debate is Andrew Shaffer, owner of the Order of St. Nick greeting card company," *Fox & Friends* host Steve Doocy said to the television audience.

It was 6:24 a.m. at the Fox News studios in New York—an early hour by anyone's watch, but it was 5:24 a.m. in Des Moines, Iowa, from where I was live via satellite. My fiancée had grilled me late into the previous night with questions that we expected a Fox News host would ask an atheist, such as "Where are your horns?" and "Why do you hate America?" It had taken a hotel wake-up call, two cell phone alarms, a Red Bull, and a gas station coffee just to pry my eyes halfway open.

Greg Stielstra, a Christian marketing expert, joined the conversation from the *Fox & Friends* set. Greg's position was that by using the secular greeting "happy holidays" in advertising and store displays instead of "merry Christmas," retailers risk alienating a majority of their customers. This wasn't semantics; this was war.

Greg: Businesses play a numbers game. They carry the most popular products. They open their stores in the busiest intersections. If 96 percent of the population is celebrating Christmas and 77 percent consider themselves Christians, why wouldn't you speak to Christmas as a retailer?

Steve: All right. Andrew, what do you make of that argument?

Me: I actually agree with that. I think that if you're trying to reach the widest possible audience, that's a great strategy.

The atheist and the Christian, finding common ground? The debate was over before it had even begun. It remains, to this day, three of the least riveting minutes of television ever produced by a major cable news channel (and that includes every episode of *Larry King Live*). It wasn't a total disaster—at least I hadn't fallen asleep on the air, something I'm pretty sure that Larry King *has* done.

While Christians are usually portrayed as the defensive side in the war on Christmas, they fired one of the first shots in 1870 when President Ulysses S. Grant signed a bill into law declaring Christmas Day a federal holiday.* If this sounds to you like a possible violation of the constitutional separation of church and state, you're not alone. Ohio lawyer Richard Ganulin sued the federal government in 1998 to have Christmas Day removed from the list of public holidays. The lawsuit was tossed out. The U.S. Court of Appeals for the Sixth Circuit, upholding a lower court's dismissal of Ganulin's lawsuit, ruled that the 1870 law does not constitute an endorsement of Christianity by the government.

Case closed, right? Not so fast. State and local governments are not required to recognize federal holidays. In the late twentieth cen-

* United States Code, Title V, Section 6103(a).

tury, city council and PTA meetings have become the de facto battlegrounds for the heart and soul of Christmas. If your children attend a public school in the United States, there is a reasonable chance they don't take two weeks off for Christmas break—it's likely they're being forced to enjoy a "holiday break" or a "winter break" instead. Christians are "asked to celebrate something they don't celebrate—winter—as if they are pagans in the Roman Empire," Fox News host John Gibson wrote in *The War on Christmas: How the Liberal Plot to Ban the Sacred Christian Holiday Is Worse Than You Thought.**

Gibson views the usage of "happy holidays" and "winter break" as evidence of a vast conspiracy to eliminate Christmas from the public sphere. The bad guys, it turns out, are not just "professional atheists"† but are, in fact, "mostly liberal white Christians." According to Gibson, the nefarious plot against Christianity is the work of ACLU lawyers, school superintendents, and city council members—many of whom are Christian—who are afraid of running afoul of the constitutional separation of church and state. They've taken the law into their own hands, rebranding Christmas trees as "friendship trees" and stopping children from handing out candy canes. One misguided soul even banned red and green decorations altogether in his school.

The Supreme Court has consistently protected expressions of Christmas on government property and in public schools. As long as a Christmas display is not entirely composed of religious symbols, for instance, court precedents point to letting things slide. This has led to

---

* Gibson was likely referring to Saturnalia, the ancient Roman festival that honored the Roman god Saturn. Beginning on December 17 and lasting a full seven days, Romans gathered with family, exchanged gifts, ate, drank, and were merry.

† I'm not exactly certain what a "professional atheist" is; however, this makes us sound as though we are incredibly organized, well-dressed, and hold meetings the first Tuesday of every month with cocktails and hors d'oeuvres.

the Supreme Court's stance being mockingly nicknamed the "three-reindeer rule"—with enough reindeer, snowmen, and elves, the religiosity of a display can be diminished to acceptable levels.

The war on Christmas isn't limited to skirmishes over the separation of church and state. Businesses are the latest grinches to enter the fray. Right-wing media had a proverbial field day with Walmart, Sears, and Target when the heathen corporations started using the term "happy holidays" instead of "Merry Christmas" in the 2000s. "I think it's all part of the secular progressive agenda . . . to get Christianity and spirituality out of the public square," Fox News' Bill O'Reilly said. "Every company in America should be on its knees thanking Jesus for being born. Without Christmas, most American businesses would be far less profitable." Conservative Christian groups now maintain lists of "naughty and nice" retailers that concerned citizens can consult to find out who's celebrating Christmas and who's celebrating "the holidays."

By December 2005, Christmas was under siege from all sides: in our schools, in our town halls, and in our most hallowed grounds (retail stores). But at least the federal government still supported Christmas.

Then the unthinkable happened: President George W. Bush and First Lady Laura Bush sent a "holiday" card.

When the biggest, baddest Christians in America dropped the H-bomb on the 1.4 million people on their Christmas card list, all hope for the future of "merry Christmas" was lost. The white flag had finally been waved. As John Lennon wrote, "War is over now / Happy ~~Xmas~~ holidays."

After the White House slight, the word *Christmas* suddenly felt dangerous and sexy. In 2007, I started Order of St. Nick, a greeting card company specializing in humorous Christmas cards. The most popular designs were a line of six tongue-in-cheek "atheist Christmas cards."

By some estimates, up to 15 percent of Americans consider themselves atheist, agnostic, or otherwise unaffiliated with any religion. If 96 percent of Americans celebrate Christmas, that means there is a large segment of the population that Hallmark and American Greetings had never spoken to directly: Santa-loving, tree-decorating, carol-singing atheists like me.

The atheist Christmas cards struck a nerve. Comedian Stephen Colbert, host of Comedy Central's *The Colbert Report,* gave me an on-air tongue-lashing that I will never forget:*

> A wag of my finger at the Order of St. Nick greeting card company.
>
> Now, I always thought that any card that was blank inside was already atheist. You open it up and see nothing but a void.
>
> Once atheists start sending Christmas cards, how long before they are including their year-end atheist family updates? . . . "Sadly, Grandpa passed away this year, but at least we know he's not in a better place. He's decomposing. Merry Xmas."

Order of St. Nick sold thousands of atheist Christmas cards after *The Colbert Report* aired. The cards didn't mock Christianity or cry for attention with cheap shock value like the drawings of Santa Claus nailed to a cross that teenage atheists doodle in their notebooks every December.† The Darwin-as-Santa image, by contrast, was a sincere expression of both my belief in atheism and Christmas'

---

* Colbert, by the way, is not an atheist. A devout Roman Catholic, he attends church and teaches Sunday school.

† This was the first Christmas card I created, back when I was a fourteen-year-old atheist. I was a very bad kid.

unique ability to bridge the gap between believers and non-believers.

For many, Christmas is already a secular holiday; believing in the Christian God is no longer a requisite for celebrating the day of His birth. The world's most famous non-believer, Richard Dawkins, exchanges gifts with his family and loves singing traditional Christmas carols. "I am perfectly happy on Christmas Day to say 'Merry Christmas' to everybody," he told Radio 4's *Today* program.

Dawkins is not alone. The Christmas season has become a time for families, regardless of religious affiliation, to get together, exchange gifts, eat cookies, and revel in "the hap-happiest time of the year." Celebrating Christmas without subscribing to Christianity is like watching the Super Bowl without having watched a single regular-season football game all year. Some people watch the Super Bowl exclusively for the commercials; others watch it for the halftime show. NFL super-fans might turn their noses up at the party crashers, but I submit that there are some spectacles so awesome that people can't help but be sucked in by their gravitational pull. Christmas sits like a black hole on the calendar, and the other holidays implied by "happy holidays"—Hanukkah, Kwanzaa, Boxing Day, New Year's Day, etc.—are powerless to be drawn in by its force. No matter how thorough the semantic cover-up job we do, we all know the Holiday Whose Name We Shall Not Speak That Is Celebrated by 96 percent of Americans Every December.

Moreover, Christmas is no longer limited to countries with Christian majorities. Christmas is beginning to show up in places where Christianity has never taken hold, such as Japan and China. Several Chinese men and women I spoke with on a trip to Guangzhou, China, a few years ago recognized Santa Claus' familiar visage . . . but they couldn't pick the Baby Jesus out of a Nativity scene. *Secular Christmas is already here*. The "church and state" court cases, the verbal wrangling over "happy holidays"—the war on Christmas that is being fought primarily in the United States, Canada, and, to an extent, the United Kingdom—feels so *unnecessary* by comparison.

Even as an atheist, I feel some of Bill O'Reilly and John Gibson's pain when the war on Christmas claims another town hall display or department store ad. Can't we just back off the Baby Jesus? Hasn't He been through enough already?* Perhaps I've just been beaten into submission. Perhaps a more passionate atheist would not concede points to a Fox News host and a faith-based marketer on national television. Perhaps I should take offense at Nativity scenes on city property; perhaps I should roll my eyes every time someone says that "Jesus is the reason for the season."

But I'm an atheist, not a vampire. I don't need to cringe at the sight of a cross in a school Christmas pageant. And Christmas has given all of us so much to be thankful for. Without Christmas, there would be no *It's a Wonderful Life,* no *Miracle on 34th Street,* no *Die Hard.* We wouldn't have Dickens' *A Christmas Carol* or Christopher Moore's *The Stupidest Angel.* And, growing up, I would never have received so many Transformers, Nintendo video games, and JC Penney sweaters. Without the warm fuzzies created in our hearts by our collective Christmas spirit and shortened workweeks, seasonal affective disorder would reach epidemic proportions every December in the Northern Hemisphere. As atheist Judith Hayes wrote, "Life is difficult and short. If we can add some merriment to it, we should go for it. Every time." So, from one atheist to another, merry Christmas.

---

* See Mel Gibson's *Passion of the Christ.*

CONTRIBUTOR BIOGRAPHIES

DAVID BADDIEL AND ARVIND ETHAN DAVID

David Baddiel is a comedian, novelist, documentary maker, and most recently screenwriter. Born and raised Jewish, and maintaining a deep affection for his Jewish heritage and identity, David's Facebook religious views entry describes him as a "fundamentalist atheist."

Arvind Ethan David is the producer behind a rash of recent British feature films, including 2009's zombie-high-school movie, *Tormented.* Born and raised Catholic, Arvind's Facebook religious views entry reads "Atheist. Humanist. Yogi. Bear."

David and Arvind are currently collaborating on *The Infidel*, a feature film about a moderate Muslim man who discovers that he was adopted and is, in fact, Jewish. The film, which stars Omid Djalili, Richard Schiff, Archie Panjabi, and Matt Lucas, was released in the UK in April 2010.

JULIAN BAGGINI

Julian Baggini (www.julianbaggini.com) is editor and co-founder of the *Philosophers' Magazine* (www.philosophersmag.com). He is the author of several books, including *Welcome to Everytown: A Journey into the English Mind* (Granta), *Complaint* (Profile), and, most recently, *Should You Judge This Book by Its Cover?* (Granta). His monthly philosophy podcast is available on iTunes or via his website. He has written for numerous newspapers and magazines, including the *Guardian*, the *Financial Times*, *Prospect*, and the *New Statesman*, and has also appeared as a character in two Alexander McCall-Smith novels.

SIÂN BERRY

Siân Berry (www.sianberry.org.uk) is a Green campaigner, writer, and politician. She was the Green Party candidate for London mayor in 2008 and helped set up the Alliance Against Urban 4×4s campaign group, famous for placing mock parking tickets on Chelsea Tractors throughout the capital and inspiring similar actions around the world.

In 2008, Siân published the 50 Ways series of four books on greener living, each giving fifty simple tips to reduce your impact on the planet and save cash in the process. Her new book, published in October 2009, makes the link between being green and saving money even clearer, and is simply called *Mend It!*

CHARLIE BROOKER

Charlie Brooker (www.guardian.co.uk/charliebrooker) is a columnist, television scriptwriter, and broadcaster. He writes two weekly columns for the *Guardian*, including "Charlie Brooker's Screen Burn," and produces, writes, and presents the satirical BBC4 shows *Charlie Brooker's Screenwipe* and *Charlie Brooker's Newswipe*. He created, wrote, and produced the zombie horror series *Dead Set* (E4,

2008) and co-wrote the Channel 4 sitcom *Nathan Barley* with Chris Morris, based on the eponymous character in his comic strip *TV Go Home.* He has just finished presenting the first series of Channel 4's *You Have Been Watching.*

ED BYRNE

Ed Byrne (www.edbyrne.com) is one of the UK's most successful stand-up comics. He has made numerous appearances on *Mock the Week*, *8 Out of 10 Cats*, *Never Mind the Buzzcocks*, Graham Norton's *The Bigger Picture*, and *Have I Got News for You*, as well as performing on several sold-out tours of the UK. Ed has appeared five times on NBC's *Late Night with Conan O'Brien*, released several DVDs including *Psychobabble* and *Pedantic and Whimsical*, hosted the Montreal Just for Laughs festival, and made his London West End debut in a two-week run at the New Ambassadors Theatre. He once starred in a *Father Ted* Christmas special as a grungy teenager.

JENNY COLGAN

Jenny Colgan (www.jennycolgan.com) is the author of nine bestselling novels, including *Amanda's Wedding* and *West End Girls.* She also contributes to a variety of publications, from *The Times* to *Cosmopolitan*, and specializes in not being asked back onto BBC panel shows. No, not even *Quote Unquote.*

Jenny lives in France with her husband and three children and likes cycling, Scrabble, stand-up comedy, and playing the piano really badly. Her latest novel, *Diamonds Are a Girl's Best Friend*, is available from Sphere, and she writes a book blog on her website.

BRIAN COX

Professor Brian Cox began his career as a musician, playing keyboards with rock band D:Ream, who went on to have many Top 10 hits including New Labour's election anthem "Things Can Only Get Better." He is now a Royal Society research fellow, a professor at the University of Manchester, and a member of the high-energy physics group at the University of Manchester, and works on the ATLAS experiment at the Large Hadron Collider, CERN, near Geneva, Switzerland. He is best known to the public as the presenter of a number of science programs for the BBC.

RICHARD DAWKINS

Professor Richard Dawkins, FRS (www.richarddawkins.net), is an internationally renowned evolutionary biologist and ethologist and the author of numerous best-selling books, including *The Selfish Gene, The Blind Watchmaker, Unweaving the Rainbow, Climbing Mount Improbable, A Devil's Chaplain, The Ancestor's Tale*, and *The God Delusion.* He was the first holder of the Simonyi Professorship for the Public Understanding of Science at Oxford, and has written and presented many television documentaries, including *Root of All Evil?*, *The Enemies of Reason*, and *The Genius of Charles Darwin.* His most recent book is *The Greatest Show on Earth* (Transworld).

NEIL DENNY

Neil Denny is the producer and co-host of the *Little Atoms* radio show (www.littleatoms.com). Created in 2005 by Neil Denny and Richard

Sanderson, and currently hosted by Neil Denny and Padraig Reidy, *Little Atoms* is a live talk show about ideas. The show is based around ideas of the Enlightenment and promotes science, freedom of expression, skepticism, and humanism, although this means we often end up discussing superstition, religious fundamentalism, censorship, and conspiracy theory. The show is broadcast on Friday evenings at seven o'clock on London's Resonance 104.4 FM. The *Little Atoms* Podcast is available on iTunes.

NICK DOODY

Nick Doody is a stand-up comic and writer. He has written for *8 Out of 10 Cats*, FAQ U, and *The Friday Night Project*, and regularly writes for *The Late Edition*, *The Now Show*, and *Armando Iannucci's Charm Offensive*. While still a student, Nick supported the late cult American comic Bill Hicks on tour, and later wrote a biography of him titled *Telling the Truth, Laughing: The Life and Works of Bill Hicks*. Nick has recently had an original series, *Bigipedia*, broadcast on Radio 4. He regularly performs comedy at secular events and supports secular causes.

EMERY EMERY

Emery Emery (www.emeryemery.com) became a stand-up comedian at the age of eighteen, touring the United States and Europe for twenty years. After editing the feature film *The Aristocrats*, Emery built a second career directing and producing for film and television.

Emery's life partner, Laney, is HIV positive. Laney and Emery are proud to be a serodiscordant couple. Through diligent safe sexual practices, Emery has remained HIV negative since becoming Laney's partner in 2005.

HERMIONE EYRE

Hermione Eyre is a journalist. She went to Rugby School but left neither muscular nor Christian. At Oxford, she interviewed Seamus Heaney for a student paper by fax machine. She was staff writer and TV critic on the *Independent on Sunday* for six years, where her most exciting interviewees were Gilbert and George, Slavoj Zizek, Jarvis Cocker, and Joanna Lumley, whom she saved from a gunman. She wrote, with William Donaldson, *The Dictionary of National Celebrity* (2005, Weidenfeld and Nicholson) and contributes to media including *New Statesman*, the *Observer*, and Radio 3's *Night Waves*.

ROBBIE FULKS

Robbie Fulks (www.robbiefulks.com) is a country songwriter, guitar picker, and recording artist. He has released nine solo albums, of which the most recent, *50-vc. Doberman*, a digital fifty-song multi-genre experiment, is available at his website. Among his proudest accomplishments are appearing at the Grand Ole Opry, producing George Jones, singing with Mavis Staples, and recording with Lucinda Williams, Sam Bush, Jenny Scheinman, Lloyd Green, and Al Anderson (though narrowing it down is hard). Robbie has lived for twenty-five years in and around Chicago.

A. C. GRAYLING

Anthony Grayling (www.acgrayling.com) is professor of philosophy at Birkbeck College, University of London, and a Supernumerary Fellow of St. Anne's College, Oxford. For several years he wrote the "Last Word" column for the *Guardian*, and he is a regular reviewer for the *Literary Review* and the *Financial Times*. He also often writes for the *Observer, Economist, Times Literary Supplement, Independent on Sunday*, and *New Statesman*, and is a frequent broadcaster on BBC Radio 4, Radio 3, and the World Service. His latest books are *Liberty in the Age of Terror: A Defence of Civil Society and Enlightenment Values* and *Ideas that Matter.*

NATALIE HAYNES

Natalie Haynes (www.nataliehaynes.com) is an award-winning comedian, writer, and broadcaster. She has written and performed five stand-up shows that have toured internationally, from Berlin to Manhattan, via the Edinburgh Fringe. She is also a classicist and the author of a forthcoming book on how the ancients shape our lives even now, *The Ancient Guide to Modern Life*, to be published by Profile Books in 2010. She is a regular contributor to the *Times*, the *Sunday Times*, and *New Humanist*, and a reviewer for *Newsnight Review* (BBC2), *Saturday Review* (Radio 4), and *Front Row* (Radio 4).

JON HOLMES

Jon Holmes (www.jonholmes.net) is a seven-time Sony Radio and British Comedy Award–winning writer, comedian, and polymorphic media alloy. He is on Radio 4 more often than the weather, co-creating *Dead Ringers* and starring in *The Now Show* as well in his own series *Listen Against*. He presents his own show on BBC6 Music and turned his first book, *Rock Star Babylon*, into a stand-up show for the 2009 Edinburgh Festival. His TV credits includes *Mock the Week* and *Have I Got News for You*, and he was once asked to go into the jungle on *I'm a Celebrity* . . . but told them to sod off.

ROBIN INCE

Robin Ince (www.robinince.com) is an award-winning comedian and writer and the creator of Nine Lessons and Carols for Godless People. He performs stand-up shows about African orchid beetles, finches' beaks, and Tycho Brahe, but is not very good at making jokes about wave particle duality. He regularly appears on Radio 4 on the likes of *The News Quiz* and *Just a Minute* and all those TV shows you would expect comedians to turn up on. He can also frequently be seen at book festivals facetiously reading from books by charlatans and bamboozlers in Robin Ince's Book Club, with a little Mills and Boon for light relief. He is currently working on impersonations of Richard Feynman.

ALLISON KILKENNY

Allison Kilkenny (www.wearecitizenradio.com and www.allisonkilkenny.com) is a journalist and radio host living in NYC. G. Gordon Liddy once said her writing makes him want to vomit, which is the greatest compliment she's ever received, ever.

JAMIE KILSTEIN

Jamie Kilstein (www.wearecitizenradio.com) is a political comedian and radio host living in Brooklyn. He is in love with one of the other contributors in this book.

MATT KIRSHEN

In 1987 Matt Kirshen (www.mattkirshen.com) first took to the stage, as King 2 in a production of the Nativity play. His performance was so immensely powerful that he could not bring himself to perform publicly for another six years, graciously giving other actors a chance. In 1993 the lure of Shakespeare was too great, and on finding out that a staging of *Macbeth* lacked a Second Servant, he boldly stepped into the breach, shamelessly stealing both scenes, before gliding back into semi-retirement. In 2001 he finally realized that the only way to stem the jealous tide of fellow performers would be to only work alone, and he embarked on a career in stand-up that continues to this day.

PAUL KRASSNER

Paul Krassner (www.paulkrassner.com) is the author of *Confessions of a Raving, Unconfined Nut: Misadventures in the Counterculture*, and publisher of the Disneyland Memorial Orgy poster. He is the only person in the world ever to receive awards from both *Playboy* (for satire) and the Feminist Party Media Workshop (for journalism). His reviews have been highly complimentary. The *New York Times*: "He is an expert at ferreting out hypocrisy and absurdism from the more solemn crannies of American culture." The *Los Angeles Times*: "He has the uncanny ability to alter your perceptions permanently." The *San Francisco Chronicle*: "Krassner is absolutely compelling. He has lived on the edge so long he gets his mail delivered there."

SIMON LE BON

Simon Le Bon is the lead singer of Duran Duran (www.duranduran.com). Since 1981, the band have had thirty UK Top 40 hits, including two number ones, and recorded twelve studio albums, three of which achieved multi-platinum status. In 1993 they were awarded a star on the Hollywood Walk of Fame, and in 2009 they headlined London's Lovebox festival. Duran Duran's latest album, *Red Carpet Massacre*, was released in 2006 on Epic Records, featuring collaborations with Justin Timberlake and producer Timbaland. Simon lives in London with his wife, Yasmin, and three daughters.

EVAN MANDERY

Evan Mandery (www.evanmandery.com) was born in Brooklyn, New York, and raised in East Meadow, Long Island. He is the author of two novels, *Dreaming of Gwen Stefani* and *First Contact (Or It's Later Than You Think)*, and two works of non-fiction. His new novel, *Q: An Unlived Memoir* will be published by HarperCollins in 2011. He is currently writing a history of the Supreme Court's treatment of the death penalty in the 1970s, to be published by Delphinium Books in 2012. Evan is a professor at the City University of New York. He is an avid poker player and golfer. He lives in Forest Hills, New York, with his wife, Valli Rajah-Mandery, a sociologist, and their three children.

ZOE MARGOLIS

Zoe Margolis is a writer and journalist, and a frequent contributor to the *Guardian* and *Observer.* Zoe's blog, Girl with a One-Track Mind, won the Bloggie Award for Best British or Irish Weblog in both 2006 and 2007, was ranked twenty-fourth on the Most Powerful Blog in the World list by the *Observer,* and was named the World's Most Famous Sex Blog by Nerve.com. The book based on the blog, also titled *Girl with a One-Track Mind,* is an international bestseller, translated into sixteen languages, and is being adapted into a screenplay.

CHRISTINA MARTIN

Christina Martin is a stand-up comic and comedy writer. She has been a regular feature writer for *Viz* since April 2006, and is the first female writer in the comic's thirty-year history. She won third place in the 2006 Funny Women Awards, and has performed live stand-up across the UK. She has also written for Radio 4's *Recorded for Training Purposes,* performed comedy for Radio 4 and BBC7, and writes regularly for *New Humanist* magazine. Christina likes playing Nintendo, monkeys, eating beef Monster Munch, and reading (books, not the place).

JENNIFER McCREIGHT

Jennifer McCreight (www.blaghag.com) graduated from Purdue University in 2010 majoring in genetics and evolution, and is currently a Ph.D. candidate at the University of Washington. During her time at Purdue, she founded the Society of Non-Theists, a student group for atheist and agnostics on campus, and acted as president for three years. She is on the board of directors for the Secular Student Alliance, and her popular atheist blog, Blag Hag, frequently covers topics such as atheism, religion, science, biology, academia, feminism, and sex.

CASPAR MELVILLE AND PAUL SIMS

Caspar Melville is editor of *New Humanist* magazine (www.newhumanist.org.uk). He formerly worked for openDemocracy. His writing has appeared in the *Los Angeles Times*, *Toronto Star*, *Sunday Telegraph*, *Village Voice*, and loads of obscure music magazines. His first book, *Taking Offence*, was published in 2009 by Seagull Books.

Paul Sims is news editor of *New Humanist* and is responsible for the magazine's online content, including the website, blog, and podcast. He graduated from Oxford University, where he studied modern history, in 2006, before moving to London to join *New Humanist* in 2007.

ANDREW MUELLER

Andrew Mueller (www.andrewmueller.net) is a journalist based in London, where he writes for *Monocle*, the *Guardian*, *Esquire*, the *Times*, *Independent*, *Uncut*, and *New Humanist*, among others. He is the author of the books *Rock and Hard Places* and *I Wouldn't Start from Here*, and appears in the index of Richard Dawkins' *The God Delusion* in between Mozart and Muhammad, which strikes him as only fitting. He also sings and plays guitar in The Blazing Zoos, the ninth-best country band in the entire E2 postcode.

GRAHAM NUNN

Graham Nunn (www.grahamnunn.net) is the official designer for the Atheist Bus Campaign and its website AtheistCampaign.org, and produced the cover layout of this book. He also designed all the Atheist Bus Campaign merchandise, which can be purchased from Blue Apple Music. In 1997, Graham refused to make the switch from postal letters to e-mail, predicting, "The Internet will destroy the world." He has now built his own computer, designed over thirty websites, learned HTML, Flash, Illustrator, and Photoshop, and, in the area of technology at least, vowed to stop worrying and enjoy his life.

PHIL PLAIT

Phil Plait is a professional astronomer, blogger, author, and president of the James Randi Educational Foundation (www.randi.org), an American non-profit organization devoted to promoting critical thinking across the world. He has dedicated his life to promoting real science and stamping out anti-science, from people who think the Apollo moon landings were faked to creationists who are trying to indoctrinate schoolchildren in the classroom. He writes about all this and more on his blog Bad Astronomy (www.badastronomy.com), hosted by *Discover* magazine (discovermagazine.com/badastronomy).

NEAL POLLACK

Neal Pollack (www.nealpollack.com) is the author of several acclaimed books of satirical fiction and nonfiction. HarperPerennial published his latest book, *Stretch: The Unlikely Making of a Yoga Dude*, in August 2010. He lives somewhere in North America with his wife and son.

SIMON PRICE

Simon Price is the rock and pop critic of the *Independent on Sunday* newspaper. He has been writing about music for twenty-five years (starting while he was still a sixteen-year-old schoolboy in South Wales), including nine years at *Melody Maker*, which he left to write the best-selling rock biography *Everything: A Book About Manic Street Preachers*. He moonlights as a club DJ and promoter, running the glam night Stay Beautiful in London and the alternative eighties night Spellbound in Brighton.

CLAIRE RAYNER

Claire Rayner is a journalist, writer, and broadcaster, and is best known as Britain's leading agony aunt. She was born in January 1931 in London and trained as a nurse, winning the gold medal for outstanding achievement when she became an SRN in 1954. Since the start of her writing career in 1960, she has written over ninety books, and in 1996 was awarded an OBE "for services to women's issues and health issues." She is a vice president of the British Humanist Association.

Claire married Desmond Rayner in 1957 and they live in North London. They have three children, three grandsons, and a granddaughter.

SID RODRIGUES

Sid Rodrigues (www.skepticsinthepub.org) works in a science laboratory, describes himself as "a lab monkey," and has worked with pipettes, test tubes, and that three-prime, five-prime stuff since 1998. He lives in South West London, which, he concedes, isn't quite as glamorous as Las Vegas, where he got married in July 2009 in front of the world's skeptical community. Still, he flies to the States quite often to see his new wife, Rebecca Watson, creator of the female skeptical site Skepchick. He loves Camembert, science books, and, of course, Rebecca.

MARTIN ROWSON

Martin Rowson is one of Britain's most acclaimed cartoonists. He has contributed to, among others, the *Guardian, New Humanist, Time Out, Independent on Sunday, Observer, Independent Magazine, Daily Mirror, Daily Express,* the *Times Educational Supplement, Tribune, Morning Star, Dublin Sunday Tribune*, and the *Scotsman*, and is the author of the book *The Dog Allusion: Gods, Pets and How to Be Human*, in which he likens pet keeping to following a religion. Martin is married with two teenage children. He lists his interests as cooking, drinking, ranting, atheism, zoos, and collecting taxidermy.

ADAM RUTHERFORD

Adam Rutherford is a professional geek. He holds a Ph.D. in genetics, works at the science magazine *Nature*, and presents radio and television programs, including *Cell* for BBC4: a series covering 4 billion years of evolution and 300 years of biology, intrigue, betrayal, and rather more sperm than is absolutely necessary. Writing for the *Guardian*'s blog Comment Is Free, his grouchy response to atheists being universally labeled as "intellectual cowards" briefly held the record for the most comments ever. He stopped claiming this as an achievement after he was soundly beaten by articles by a Radio 1 DJ, a posh gap year student, and one about that godforsaken bus.

ANDREW SHAFFER

Andrew Shaffer is the creative director of Order of St. Nick (www.orderofstnick.com), the greeting card company whose irreverent cards have been featured on a variety of popular media outlets including *The Colbert Report*, the *Washington Post*, and NPR. He has a graduate degree from the University of Iowa, where he also attended the Iowa Writers' Workshop for a summer semester. He lives in Iowa with his wife. Shaffer's first non-fiction book is *Great Philosophers Who Failed at Love* (HarperPerennial, 2011).

SIMON SINGH

Simon Singh (www.simonsingh.net) is an author, journalist, and director, who specializes in science and mathematics. He joined the BBC after completing a Ph.D. in particle physics at Cambridge University. He worked on *Tomorrow's World* and then directed a documentary for the *Horizon* series on the subject of Andrew Wiles and his proof of Fermat's last theorem, which won a BAFTA in 1997. He went on to write a book titled *Fermat's Last Theorem*, and is also the author of *Big Bang* and *The Code Book*. His latest book is *Trick or Treatment?*

Alternative Medicine on Trial, which he co-authored with Edzard Ernst, the world's first professor of complementary medicine.

DAVID STUBBS

David Stubbs writes about music, TV, and sport for, among others, the *Guardian*, *The Sunday Times*, *The Wire*, *Uncut*, and *When Saturday Comes.* He was formerly a staff writer at *NME* and *Melody Maker*, for whom he created the Mr. Agreeable character (www.mr-agreeable.net). His most recent book is *Fear of Music: Why People Get Rothko but Don't Get Stockhausen* from zerO books, which is all about why people get Rothko but don't get Stockhausen. He had a very good Catholic upbringing. He currently lives in Woolwich with his conscience.

CATIE WILKINS

Catie Wilkins is a comedy writer and stand-up comic. She's gigged all over the UK at clubs including the Comedy Store, Comedy Café, The Stand, Banana Cabaret, Up the Creek, and The Komedia; appeared on the DVD of *Jimmy Carr's Comedy Idol*; and performed in two Edinburgh shows, *Comedy o'Clock* (2008) and *It's Got Jokes In* (2009). She has written sketches for The Works at Madame Jojo's, and has a sitcom, *Nothing New Under the Sun*, in development with BBC3. Catie was born in 1980 and lives in London.

ABOUT THE EDITORS

ROBIN HARVIE

Robin Harvie is an atheist, publisher and author of *Why We Run: The Story of an Obsession*. He lives in London, UK.

STEPHANIE MEYERS

Stephanie Meyers is a book editor and an avid believer in receiving presents, decorating trees, and making the most of post-holiday sales. She lives in New York City.

SECULAR RESOURCES

American Atheists (www.atheists.org) is a nationwide movement that defends civil rights for atheists; works for the total separation of church and state; and addresses issues of First Amendment public policy.

The **American Humanist Association** (www.americanhumanist.org) advocates for the rights and viewpoints of humanists, striving to bring about a progressive society where being "good without god" is an accepted way to live life.

The **Foundation Beyond Belief** (www.foundationbeyondbelief.org) was created to focus, encourage, and demonstrate the generosity and compassion of atheists and humanists, and to provide support and encouragement for non-theistic parents.

The **Freedom from Religion Foundation** (www.ffrf.org) is an educational group working to promote the constitutional principle of separation of state and church, and to educate the public on matters relating to non-theism.

The **Secular Coalition for America** (www.secular.org) is an advocacy organization whose purpose is to amplify the diverse and growing voice of the non-theistic community in the United States.